TK
7870
P5
S25
1987

LIBRARY

JUN 0 5 1987

FAIRCHILD R & D

Encapsulation of Electronic Devices and Components

ELECTRICAL ENGINEERING AND ELECTRONICS

A Series of Reference Books and Textbooks

Editors

Marlin O. Thurston
Department of Electrical
Engineering
The Ohio State University
Columbus, Ohio

William Middendorf
Department of Electrical
and Computer Engineering
University of Cincinnati
Cincinnati, Ohio

1. Rational Fault Analysis, *edited by Richard Saeks and S. R. Liberty*

2. Nonparametric Methods in Communications, *edited by P. Papantoni-Kazakos and Dimitri Kazakos*

3. Interactive Pattern Recognition, *Yi-tzuu Chien*

4. Solid-State Electronics, *Lawrence E. Murr*

5. Electronic, Magnetic, and Thermal Properties of Solid Materials, *Klaus Schröder*

6. Magnetic-Bubble Memory Technology, *Hsu Chang*

7. Transformer and Inductor Design Handbook, *Colonel Wm. T. McLyman*

8. Electromagnetics: Classical and Modern Theory and Applications, *Samuel Seely and Alexander D. Poularikas*

9. One-Dimensional Digital Signal Processing, *Chi-Tsong Chen*

10. Interconnected Dynamical Systems, *Raymond A. DeCarlo and Richard Saeks*

11. Modern Digital Control Systems, *Raymond G. Jacquot*

12. Hybrid Circuit Design and Manufacture, *Roydn D. Jones*

13. Magnetic Core Selection for Transformers and Inductors: A User's Guide to Practice and Specification, *Colonel Wm. T. McLyman*

14. Static and Rotating Electromagnetic Devices, *Richard H. Engelmann*

15. Energy-Efficient Electric Motors: Selection and Application, *John C. Andreas*

16. Electromagnetic Compossibility, *Heinz M. Schlicke*

17. Electronics: Models, Analysis, and Systems, *James G. Gottling*

18. Digital Filter Design Handbook, *Fred J. Taylor*

19. Multivariable Control: An Introduction, *P. K. Sinha*

20. Flexible Circuits: Design and Applications, *Steve Gurley, with contributions by Carl A. Edstrom, Jr., Ray D. Greenway, and William P. Kelly*

21. Circuit Interruption: Theory and Techniques, *Thomas E. Browne, Jr.*

22. Switch Mode Power Conversion: Basic Theory and Design, *K. Kit Sum*

23. Pattern Recognition: Applications to Large Data-Set Problems, *Sing-Tze Bow*

24. Custom-Specific Integrated Circuits: Design and Fabrication, *Stanley L. Hurst*

25. Digital Circuits: Logic and Design, *Ronald C. Emery*

26. Large-Scale Control Systems: Theories and Techniques, *Magdi S. Mahmoud, Mohamed F. Hassan, and Mohamed G. Darwish*

27. Microprocessor Software Project Management, *Eli T. Fathi and Cedric V. W. Armstrong (Sponsored by Ontario Centre for Microelectronics)*

28. Low Frequency Electromagnetic Design, *Michael P. Perry*

29. Multidimensional Systems: Techniques and Applications, *edited by Spyros G. Tzafestas*

30. AC Motors for High-Performance Applications: Analysis and Control, *Sakae Yamamura*

31. Ceramic Materials for Electronics: Processing, Properties, and Applications, *edited by Relva C. Buchanan*

32. Microcomputer Bus Structures and Bus Interface Design, *Arthur L. Dexter*

33. End User's Guide to Innovative Flexible Circuit Packaging, *Jay J. Miniet*

34. Reliability Engineering for Electronic Design, *Norman B. Fuqua*

35. Design Fundamentals for Low-Voltage Distribution and Control, *Frank W. Kussy and Jack L. Warren*

36. Encapsulation of Electronic Devices and Components, *Edward R. Salmon*

Additional Volumes in Preparation

Electrical Engineering-Electronics Software

1. Transformer and Inductor Design Software for the IBM PC,
   Colonel Wm. T. McLyman

2. Transformer and Inductor Design Software for the Macintosh,
   Colonel Wm. T. McLyman

3. Digital Filter Design Software for the IBM PC,
   Fred J. Taylor and Thanos Stouraitis

FAIRCHILD RESEARCH
TECHNICAL INFORMATION CENTER
4001 Miranda Ave. Dept. 3230
Palo Alto, CA 94304
(415) 855 4226

Encapsulation of Electronic Devices and Components

Edward R. Salmon

Electronic Materials, Inc.
New Milford, Connecticut

Marcel Dekker, Inc.   New York and Basel

RECID = 10664 - 1

Library of Congress Cataloging-in-Publication Data

Salmon, Edward R.
  Encapsulation of electronic devices and components

  (Electrical engineering and electronics; 36)
  Includes index
  1. Electronic apparatus and appliances--Plastic
embedment. I. Title. II. Series.
TK7871.15.P5S25 1987   621.381'046   86-19886
ISBN 0-8247-7589-9

Copyright © 1987 by Marcel Dekker, Inc. All rights reserved

Neither this book nor any part may be reproduced or transmitted
in any form or by any means, electronic or mechanical, including
photocopying, microfilming, and recording, or by any information
storage and retrieval system, without permission in writing from
the publisher.

MARCEL DEKKER, INC.
270 Madison Avenue, New York, New York 10016

Current printing (last digit):
10 9 8 7 6 5 4 3 2 1

PRINTED IN THE UNITED STATES OF AMERICA

Preface

Before encapsulating an electronic device or component, important choices are made concerning the material and method to be used. Information is necessary to make the best choices, and during my twenty-five years of dealing with thermoset polymers as both a resin supplier to formulators and major electrical end-users, and as a formulator dealing directly with electrical and electronics insulation users, I have seen that this information is not easily available. The same questions are asked. The same choices are offered: epoxy versus silicone, potting versus coating, molding versus casting. But the information necessary in selecting materials and application methods was lacking.

The electrical engineer, mechanical engineer, technician and buyer needed a materials chemist to explain the trade-offs in any encapsulation choice. They needed to know the right questions to ask and to have sufficient background that they could understand the answers they received.

The solution to this problem was to write a book, which I have done hoping this accumulated knowledge, necessary to making the right decisions, can be shared broadly throughout the industry.

Edward R. Salmon

Contents

Preface   *iii*
Introduction   *viii*

1. **Practical Definitions**   1
    I.   Introduction   1
    II. Definitions   2
    III. Terms and Symbols   13
    IV. Summary of Cured Encapsulant Hardnesses   14
    V.   Important Reactive Groups   15
    VI. Electrical Temperature Classifications   16
    VII. Viscosity Guide   16
    VIII. Important Abbreviations   17

2. **Class I Materials**   **19**
    I.   Epoxies   19
    II. Silicones   37
    III. Urethanes   42
    IV. Unsaturated Polyesters   47
    V. Other Liquid Encapsulating Polymers   49
    Appendix: Sources for Materials   53

3. **Class II and III Materials**   **57**
    I.   Class II Materials   57
    II. Class III Materials   65
    III. Parylene   73
    Appendix: Sources for Materials   74

4. **Methods of Encapsulation**   **77**
    I.   Introduction   77
    II. Class I Materials   78
    III. Class II Materials   112
    IV. Class III Materials   126
    V. Parylene   129
    References   132
    Appendix: Sources for Materials   133

5. **Material Selection**   **137**
    I.   Steps in Selecting an Encapsulant   137
    II. Selecting the Materials   139
    References   172
    Appendix: Sources for Materials   173

6. **Preparing to Encapsulate (by Tom Baker)**   **177**
    I.   Introduction   177
    II. Preparing to Encapsulate   178
    III. Encapsulating   184
    IV. Some Other Considerations   188

7. **Troubleshooting**   **191**
    I.   Introduction   191

II.   Class I Materials   192
III. Class II Materials   201
IV.   Class III Materials   203
V.   Parylene   206
References   206

8. **Decapsulation:**   **207**
    I.   Introduction   207
    II. Class I Materials   208
    III. Class II Materials: Repairs   214
    IV. Class III Materials   215
    V.   Parylene   219
    References   219

Index   *221*

Introduction

Why encapsulate? The reasons are nearly as numerous as the types of devices encapsulated:

> To protect the device from the environment
> For security—to hide the circuitry
> To isolate the components electrically from each other
> To hold things together
> To get heat out
> To keep moisture out
> To prevent preset adjustments from being changed
> To shockproof the device
> To mount the device
> Because the specification says so

These are only a few of the reasons. Often the main reason is just to keep dust, moisture, and fingers out, as many of the devices will operate "open-frame," that is, in air, in the absence of an encapsulant. For whatever the reason, encapsulation is done many thousands of times every day.

The materials available to accomplish this encapsulation are also nearly as numerous as the devices to be encapsulated.

A tremendous number of liquid and solid plastic and natural materials have been used to cover and protect electrical and electronic components and devices, but not all are of commercial significance today. A list of "have been used" materials would include tar, pitch, wax, cement (tried by a few), polyesters, epoxies, silicones, urethanes, acrylic solutions, varnishes, polyamide (solutions), and phenolics, alkyds, and diallyl phthallate (DAP)—these latter three being solids before, as well as after, application.

From the above list, epoxies, silicones, and urethanes represent (by far) the largest tonnage consumed in electrical protections, with the acrylics, phenolics, and DAPs being much smaller, but significant in amount used.

The list includes liquid and solid encapsulants, as well as materials in a solvent. The way the materials are applied, and the job demanded of them, often dictates the type of material that must be used and its form. These group into three classifications, each quite different in the type and/or form of the material used.

Class I Materials (Liquid potting, casting, encapsulation systems) are 100% solid liquids (including pasty dipping compounds) and convert (cure irreversibly) to thermoset solids. No material is given off or lost during this process.

Class II Materials (Compression transfer and injection molding powders) are solid materials (usually powdered) prior to application, which then proceed through a transient liquid state when heated, then cure back to a solid state, this time as an irreversible thermoset solid.

Class III Materials (Conformal coatings): Before application, they are usually solids or viscous liquids in a solvent, which are then applied as a thin film. The solvent subsequently evaporates. In some cases, such as with acrylics, that is the end of the process. The plastic protective film is still a thermoplastic, however, and

can be readily redissolved. Other solids, such as epoxies and urethanes, subsequently cure (irreversibly) to a thermoset solid. In
either case, the deposited film is only a fraction (sometimes as low
as 25%) of the volume of the original solution, due to the evaporation of the carrier solvent.

Obviously, only Class I and II materials allow for mass casting,
with the solvent diluted materials being restricted to thin film
coatings, usually less than 10 mils thick.

The requirements in material, form, and equipment for each
of these classes is so different that each one will be studied
separately.

Epoxies and silicones are used in all three classes, with urethanes in I and II, and acrylics in Class III material only.

Because the materials and methods of Class I predominate
in encapsulation, proportionally more time will be spent on them.
Chapter 2 will be devoted exclusively to Class I materials, with
Chapter 3 covering Classes II and III.

Considerable detail will be supplied on the chemistry and
identity of curing agents used to cure epoxies. Just as Class I materials dominate in encapsulation of electronics, so do epoxies
dominate the materials used in Class I. Not only is more epoxy
used than silicone and urethane combined, but when encapsulation is first considered, epoxies are usually the material given first
consideration.

Many epoxy systems are supplied as a matched pair: part A
to be used with part B. But many others are supplied with a choice
of curing agent. Many of these curing agents are well-known reactive chemicals, and when used are often known and identified as
such, allowing the user in many cases to adjust such properties as
viscosity, cure time, pot life, hardness, and so on, by curing agent
selection. Therefore, considerable time will be spent describing the
various types of epoxy curing agents available, and their effect on
cured epoxy properties.

This is in contrast to urethanes, which are always supplied as
"matched pairs," part A and part B, with no freedom to select
other curing agents. Consequently, less time will be spent on urethane chemistry.

RTV silicones offer some limited freedom to alter cure rate, rather than end properties.

The methods used to encapsulate in terms of numbers fall only a little short of the number of materials available to accomplish the encapsulation. The method is often dictated by the type of material considered necessary for the device; but method should be very much a part of the material selection process from the beginning.

The methods used vary from absurdly simple (just pour it in or dip the device in it), to the expensive (transfer or injection molding, which can exceed a quarter of a million dollars for a press and molds), to the very sophisticated. Parylene requires a very hard vacuum and pyrolysis of a solid plastic.

Chapter 1 consists of definitions, the idea being to learn a few terms first, then use them in the subsequent chapters. Chapters 2 and 3, as previously explained, cover encapsulation materials, the major ones in considerable detail. Chapter 4 explores the many methods used to apply the encapsulants of Chapters 2 and 3. Where possible, schematic diagrams of equipment are included as well as, in many cases, pictures of commercial units. Chapter 5 discusses selection of the right material for the encapsulation job. It cannot make the material decision for you since each such situation is unique to the particular device and circumstances at hand. Rather, it attempts to give ideas, trade-offs, potential problems, and so on, to help in the selection process.

Chapter 6 was written by Tom Baker, Principal Engineer, Materials Engineering Section, Equipment Development Laboratories, Equipment Division, Raytheon Corporation, Sudbury, Massachusetts. Tom brings to bear in writing this chapter nearly thirty years of materials and encapsulation experience, twenty-three of them with Raytheon. It is devoted exclusively to the use of liquid encapsulants because the most sophisticated and demanding devices must be protected by the use of these types of materials. Their intricacy, size and end-use demands often involve airborne or even space environments, and the delicate nature of most, if not all, of their individual components demands greater protection than that afforded by conformal coatings. On the other hand, the devices could not stand the rigors or temperatures encountered in molding the powders of Class III, even if they could be handled by the molding equipment. Tom takes the reader through the critical "preparing to encapsulate" stage, then through the equally critical actual steps in the encapsulation process.

Chapter 7 covers encapsulation problems and their probable or most likely causes, then gives some hints on overcoming them. The list is, of course, not exhaustive. It could never be, as new problems are invented daily, it seems. The problems listed in Chapter 7 do seem to repeat often, so the reader will, to some small degree, be forearmed.

Finally, Chapter 8—how do you get it out, if you need to, once it's been encapsulated? This chapter is rather short since decapsulation, with one exception, varies from difficult to very nearly impossible. Some difficult decisions must be made early in the material selection process if easy decapsulation on a regular production or field return basis is necessary. Make sure it's "right" before you encapsulate it.

1
Practical Definitions

I. INTRODUCTION

Ground rules and definitions are necessary to the understanding of any game or job. Although some of the following definitions are not "dictionary exact" and variations may be found in plastics dictionaries and encyclopedias, they *are* in use in the industry and will be understood by those in the "fraternity." In a few cases, a word or phrase is sometimes used to describe more than one method, phenomenon, and so on, the meaning being specific to the user. In these cases one definition, consistent with majority usage in the industry, has been established and used consistently throughout. Before alphabetizing and defining, we should look at the word "encapsulation" and establish what it means versus several apparent synonyms, at least with regard to the way these methods will be defined and used here.

II. DEFINITIONS

A. Methods for Protecting Electronics

The following five methods of protecting electronics will be used throughout the balance of this book as defined below.

Encapsulate: Commonly used two ways: (1) as a "generic" term to describe the protection of an electrical or electronic device with a plastic material, as in the title; and (2) as the forming of a thick protective envelope (from 10 to 200 mils thick) around a device or component, usually applied by a dip process, sometimes "buttered" or "spatuled" on.

Pot: The placing of a device or component in a container (the "pot"), usually referred to as a "can" if metal, or a "shell" if plastic. The container is then filled with the potting material. The container thus becomes the exterior wear surface of the component or device.

Cast: Same as pot, except that the container is a temporary mold which is subsequently removed and is usually reused. The insulation (potting material) thus becomes the exterior wear surface of the component or device. The only difference between a cast and a pot is removal of the container. The word "molded" is sometimes used to describe a component protected this way, but for clarity, the phrase "liquid molded" should be used.

Conformal Coat: Application of a thin coating, usually 10 mils or less, most often over a completed printed circuit (PC) board. Provides some environmental protection and fixturing of components on the board. Conformal coatings are applied by spray, brush, or dip.

Mold (verb): Placing a component or device in a two-piece metal mold that is clamped. A hot molten plastic is forced in under pressure. When cured, the mold is opened and the molded component or device is ejected. The cycling times for molds are much faster than those for casts.

B. Other Definitions

Accelerator: An additive to an insulation system to speed up (accelerate) its rate of reaction (cure). Does not react with either the base or the reactor portion but does remain in the system. *Synonym*: kicker.

Adduct: A prereacted form of a very reactive material—to make it less reactive and/or more easily handled. Ex-amines are partially prereacted with some epoxy. The resultant "adduct" is still highly amine reactive but is now less volatile than the pure amine and has a higher ratio and is therefore more easily measured. It gives the same cured properties. *See also* Prepolymer.

Aliphatic: Structural description of a resin or curing agent; means straight chain as opposed to a circular structure. The structure affects the reactivity. Aliphatic amines are room-temperature curing with performance to 130°C. Known as skin sensitizers. Aliphatic epoxies react very slowly at room temperature. *Examples of aliphatic structure*:

$$H_2 N-CH_2-CH_2-NH-CH_2-CH_2-NH_2$$

(Amine)

$$\overset{\displaystyle O}{\overset{\displaystyle /\ \backslash}{H_2 C-\underset{\underset{H}{|}}{C}-\underset{\underset{H_2}{|}}{C}-O[CH_2-CH_2 O]_n}} \overset{\displaystyle O}{\overset{\displaystyle /\ \backslash}{\underset{\underset{H_2}{|}}{C}-\underset{\underset{H}{|}}{C}-CH_2}}$$

(Epoxy)

Amine: A very reactive nitrogen-containing compound used to cure urethanes and epoxies. There are three forms: aliphatic, cycloaliphatic, and aromatic.

Anhydride: A "condensed" state of a dicarboxylic acid (i.e., two organic acid groups). Used as an epoxy curing agent. Offer improved heat stability over amine-cured epoxies. Require higher temperatures and longer times to cure than do aromatic amine curing agents. Usually used with an accelerator to bring cure times and temperatures down to realistic levels. Offer very long pot life (days). *Synonym*: acid anhydride.

Aromatic: A cyclic structure, usually more stable than aliphatic. The most heat stable urethane and epoxy amine curing agents are aromatic amines. Require heat when curing epoxies and offer long pot life in epoxies (hours). *Example of an aromatic amine*:

ANILINE

Synonym: heat-cured amine.

Base: The larger-volume portion of a two-component encapsulant to which the curing agent is added. Usually contains the epoxy or the isocyanate prepolymer. *Synonyms*: part A; resin.

Blush: A greasy, tacky-feeling coating on the surface of aliphatic amine–cured epoxies. The amine at the surface reacts with atmospheric moisture and carbon dioxide. Not to be confused with soft, incompletely cured materials. Aggravated by high humidity and low temperatures. *Synonyms*: spew; exudation; carbonation.

Can: A metal container used to pot a device. Remains and becomes a part of the device. *See also* Pot (Section II.A).

Carbonyl: A group of two atoms—carbon and oxygen—in acid anhydride epoxy curing agents.

Cast: See the definition in Section II.A.

Catalyst: A "magic" ingredient in many epoxy, urethane, and room-temperature vulcanizing (RTV) silicone materials. A catalyst allows a reaction to occur, which would not normally take place— at least without the destructive use of high temperatures or a very, very long wait. Very small amounts (parts per million to 1/2%) are used. A catalyst does not become a part of the cured system and is theoretically recoverable unchanged. The word "catalyst" is often used *incorrectly* to describe epoxy curing agents.

Cavity: The hole in a mold into which the device is placed prior to filling with the encapsulant.

Charge (noun): The amount of powder used in the transfer or compression molding machine.

Complex (noun): A very reactive material is loosely attached to another material to inactivate it. The reactive material can easily be displaced when desired, by heat or a more reactive material. *Example*: boron trifluouride monoethylamine (BF_3-MEA), an epoxy curing agent. Because it is inactive at room temperature, it can be added to an epoxy, thus creating a one-component system, which is subsequently cured with heat.

Compression Molding: A powder is placed in the cavity of a two-piece metal mold, which is placed on a press, clamped shut, and heated. The powder softens and flows, assuming the shape of the cavity. Process time: 2 to 4 minutes. Very capital intensive.

Conformal Coating: A thin (10 mils or less) coating, usually clear, which is applied over a PC board or component. Provides environmental protection and some fixturing. Conforms to the irregularities of the components on the board.

Coreactant: The second reactive component in a two-component system. *Example*: an amine is a coreactant with the epoxy resin.

Cross-link: The mechanism of cure. Chains literally cross-connect until the liquid material becomes a high-molecular-weight infusible solid. Cross-linking is irreversible under normal conditions.

Cross-linker: *See* Curing agent.

Cure: The hardening of the encapsulant after it is applied to the device. *Synonyms*: cross-link; gel.

Cure Time: The time for the encapsulant to harden to the point where the protected device can be handled and tested. *Synonyms*: turnover time; setup time.

Curing Agent: A material with multiple reactive points, which reacts with epoxies and polyurethanes to harden or cure them. Each reactive point on a curing agent reacts with one reactive point on the epoxy or urethane. These 1:1 matchups must be maintained. A curing agent *is not* a catalyst, although it is frequently referred to as such. *Synonyms*: hardener; reactor; part B; cross-linker.

Cycloaliphatic: A cyclic structure in which an aliphatic structure has been cyclized (i.e., the two ends attached). More heat stable and slower reacting than aliphatic. *Example of cycloaliphatic amine*:

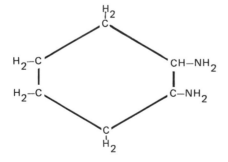

DIAMINO CYCLOHEXANE

Defoamer: An additive that aids in air bubble release in liquid encapsulants.

Dermatitis: A skin reaction (similar to poison ivy) to some of the curing agents and resins used in encapsulants. Aliphatic and cyclo-aliphatic amines are the biggest offenders.

Diluent: A low-viscosity liquid added to a liquid encapsulant to lower the viscosity. There are two types:

> *Nonreactive*:   acts as a viscosity reducer, but does not become a part of the cross-linking system. Becomes immobilized by intermingling with the chains. Significantly reduces hardness, rigidity, and softening point.
>
> *Reactive*:   as above, but also reacts with the resin system. Does not significantly reduce hardness or rigidity and lowers the softening point less than do the nonreactives.

Dimer: Two molecules of a reactive material linked together.

Dimer Acid: Two long-chain fatty acids linked together. Used in, and as, epoxy curing agents.

Draft: A slight taper to the sides of a mold to facilitate removal of a casting.

Encapsulant: The material used to encapsulate (pot, cast, mold, coat) the device.

Encapsulate: See the definition in Section II.A.

Eutectic: A liquid at room temperature composed of two materials, each of which is solid at room temperature. When melted together in the correct proportion, a liquid eutectic results. *Example*: the liquid aromatic amine used in epoxy curing agents.

Exotherm: The heat produced as a consequence of the cross-linking (curing) reaction of the encapsulant.

Filled Encapsulant: An encapsulant containing fillers, as opposed to one containing none. *Synonym*: filled system.

Filler: A dry powder, usually mineral based, used in encapsulants to improve physical, electrical, and handling properties, and to lower cost.

Flash: Material that oozes through the seams of a two-piece mold during the molding and curing process. It must be removed and is waste.

Flexible Encapsulant: An encapsulant that when fully cured displays an A-scale hardness of 80 or less. *See also* Semiflexible; Semirigid; Rigid.

Fluid Bed: A controlled "sandstorm" of a dry, powdered encapsulant, produced by low-pressure air coming upward through a cylinder containing the dry powder. A component to be coated is preheated and dipped into it, and is thus coated. Frequently, postcure is necessary.

Fluxed: The melting together of the several solid ingredients of a fluid bed or molding powder. This melting must be done at a temperature below that necessary to initiate cure. The fluxed material is subsequently cooled, crushed, and classified.

Functional-Reactive: As used to describe a reactive group. *Example*: The functional group in a polyurethane material is the isocyanate group.

Functionality: The number of reactive groups on a molecule. *Example*: The functionality of most liquid epoxies is 2 (two epoxy groups per molecule). *Synonym*: reactive point.

Gate: The entry point into the mold for the molten plastic in transfer and injection molds.

Gel (noun): A very soft encapsulant, usually A-scale hardness of 30 or less.

Gel (verb): The process of a liquid encapsulant becoming a solid but before it is fully hardened and cured. *Synonyms*: harden; set up.

Gel Time: The time (measured) from the instant of mixing until the encapsulant has gelled.

Glycol: An aliphatic molecule with a hydroxyl group at each end. The backbone chain is composed of alternating organic groups connected by an oxygen atom. *Example*:

$$HO-CH_2-CH_2-O-CH_2-CH_2-OH$$

Diethylene glycol

Hardener: *Synonyms*: curing agent; reactor; part B; cross-linker.

Hydrocarbon: An organic molecule whose backbone chain is composed of a chain (two or more) of carbon atoms. *Example*:

$$CH_3 CH_2 CH_2 CH_2 CH_2 CH_3 - hexane = CH_3 - (CH_2)_4 - CH_3$$

Induction Period: A period of time allowed between mixing and application of certain epoxies and their amine-type curing agents. This time allows the two ingredients to prereact a little and acts to lessen or eliminate the sticky surface found on some cured epoxies under certain conditions. *Synonym*: prereaction time.

Inhibition: The stopping or slowing of the cure of an encapsulant by a contaminating material that is hostile to the curing mechanism. Most often encountered in silicone and polyester systems. *Synonym*: poisoning.

Injection Molding: Like transfer molding except that the powder is fed from a hopper into a heated, jacketed screw, which turns and "injects" it into the mold cavity.

Latent: Inactive at room temperature. Becomes active when temperature is raised or other exceptional conditions are imposed. *Example*: BF_3-MEA is a latent epoxy catalyst. Becomes active when the temperature is raised above 200°F.

MEK Peroxide: Methyl ethyl ketone peroxide. The most commonly used catalyst used to initiate the cure of unsaturated polyester materials.

Moisture Resistance: Resistance of a cured encapsulant to loss of electrical and physical properties upon exposure to humidity.

Moisture Sensitivity: Tendency of one or both components of an encapsulant to react with atmospheric moisture (undesirable), resulting in degraded or complete loss of cured properties.

Mold (noun):   A block of material containing a cavity used to coat or mold a device.

Mold (verb):   See the definition in Section II.A.

Mold Gate: *See* Gate.

Mold Release: A substance coated on the interior surface of a mold to facilitate removal of a casting or molded device once the encapsulant has hardened.

Mold Vent: Exit points in a mold to permit release of displaced air and gases, if any, during molding. Vents are very critical to proper injection and transfer mold design.

Molecule: A group of atoms connected together to form a substance.

Monomer: A single reactive unit. Usually refers to the low-viscosity reactive diluent used in polyesters. *Example*: styrene.

Oligomer: A short-chain polymer.

One-Component Systems: An encapsulant in which the base and a special curing agent that displays no (or very low) room-temperature reactivity are premixed and are stable. Application of heat effects cure. The room-temperature stability of a one-component systems varies from days to months. *Synonyms*: one-can system; one-package system; one-part system.

Part B: The reactor (curing agent) portion of a two-component encapsulant. *Synonyms*: hardener; cross-linker; reactor.

Parylene:*   A special conformal coating based on polyparaxylylene. Very thin (< 1 mil) and extremely uniform pinhole-free coatings are possible. Special equipment is necessary to apply it.

PHR: Parts per hundred of resin. Used to describe the parts of curing agent used per 100 parts of the reactor portion of a two-component encapsulant.

Pigment: A colorant used in encapsulants to opacify and color the cured material.

_____

*Trademark of Union Carbide.

Platinum: A type of catalyst used in the reactor portion of some RTV silicones.

Polyamide: An amine-reactive curing agent used principally in epoxies. Gives long pot life and tough, rather than brittle, cured systems. Least potential for dermatitis of all the amine types.

Polyether: Term used to describe the backbone chain composition of certain materials. *Example*: Polyether glycols are used to cure urethane materials and modify some epoxies. Some flexible epoxies have polyether backbones.

Polymer: A long line of atoms connected to form a chain.

Polyol: An aliphatic material containing two or more hydroxyl (OH) groups.

Postcure: A heat treatment used to develop more fully the physical and/or the electrical properties of a gelled encapsulant. Performed after the encapsulant has hardened.

Pot (verb): See the definition in Section II.A.

Pot Life: The time after the two components of an encapsulant are mixed, during which the mixture can still be applied to the device. The pot life of a mixture is past when it has reacted to the point that it will not flow sufficiently to be applied. *Synonyms*: working life; working time; usable life; open time.

Potting Shell: A plastic container in which a device or component is potted.

PPM: Parts per million.

Preform: A compacted pellet of molding powder. Permits easier handling of the powder and quick, precise control of the amount charged to the press. Also facilitates preheating of the charge where necessary.

Prepolymer: A prereacted form of a very reactive material to make it less reactive (and more usable). Example: Diisocyanantes are prereacted with amines or glycols to a higher molecular weight but with residual isocyanate groups present. *Synonyms*: oligomer; adduct. Prepolymer usually refers to isocyanate materials. *See also* adduct.

Primer: A coating applied to a surface to enhance the adhesion of a subsequent encapsulant. Most often used with RTV silicones.

Pyrolysis: Heating a material to a temperature above its stability point. Usually results in destruction of the material.

Reactive Point: Reactive group of atoms in resins and curing agents that permit them to combine and cross-link to a cured state. *Synonym*: functional group. *Example*: Epoxy and isocyanate groups are reactive points.

Reactor: *Synonyms*: part B; hardener; curing agent; cross-linker.

Resin: A plastic polymer. *Examples*: epoxy; urethane. Also used to mean the base portion of a two-component encapsulant.

Reversion: The "unpolymerization" or cure reversal of certain types of encapsulants. Condensation-cure RTV silicones and polyurethanes can exhibit this phenomenon under certain conditions of high humidity and/or high temperature.

Rigid: A very hard state of cure of an encapsulant. Greater than 80 D-scale hardness. Very inflexible.

RTV: Room-temperature vulcanizing.

Runners: Canals connecting the heating chamber and the mold cavities in a transfer or injection mold. The liquefied powder travels through these canals. Residual material in the runners is lost and becomes a part of the waste.

Semiflexible: A cured encapsulant with hardness between A scale 80 and D scale 50.

Semirigid: A cured encapsulant with a D-scale hardness greater than 50 but less than 80.

Shell: *See* Potting Shell.

Sintering Temperature: The temperature at which powder particles begin to soften and become cohesive.

Spiral Flow: A two-piece metal mold with a standard spiral-shaped cavity. It looks like a flattened conical coil spring. It is used to judge the flow properties of thermoset powders and to assess the age and utility of powders in inventory. The distance the molten material flows through the spiral before it gels at a predetermined temperature and clamp pressure is measured and reported in inches.

Thermoplastic: A plastic material that can be repeatedly formed and then reformed each time it is reheated. *Examples*: polyethylene; plexiglass.

Thermosetplastic: A plastic that cures and sets. The cure is irreversible and a thermoset cannot be reformed by heating. *Examples*: epoxies; silicones; phenolics.

Tin Soap: A catalyst used with certain RTV silicones and urethanes. *Example*: dibutyltin dilaurate (DBT).

Transfer Molding: A process in which a powdered encapsulant is heated in a chamber to the liquid point, then pushed by a piston through runners to one or more cavities in a clamped metal mold where it hardens (cures). Process time: 2 to 5 min. Very capital intensive.

Unsaturated: Two carbon atoms connected by two bonds rather than the normal one [i.e., C=C (double bond)]. A very reactive group toward certain materials. *Example*: peroxides. This is the functional group that crosslinks polyesters and addition-cure RTV silicones.

Urethanes: *General Definition*: Refers to the product of an isocyanate reacted with any curing agent.

   Chemical Definition: The reaction product of an isocyanate and a hydroxyl-containing material. The reaction product of an isocyanate and an amine is a urea.

III. TERMS AND SYMBOLS

B	Boron, a metallic element
C	Carbon
Cl	Chlorine
$-CH_3$	Methyl

CO_2	Carbon dioxide
F	Fluorine
H	Hydrogen
N	Nitrogen
NH_3	Ammonia
O	Oxygen
Pt	Platinum
R	When attached to a polymer chain, denotes a group attached here; usually, the group it stands for is described nearby. (*Example*: $R = -CH_2-CH_3$)
Si	Silicon
Sn	Tin

Phenyl

Δ	Heat, added to or developed by a reaction (cure); a + or - sign denotes heat given off or required
↑	Something given off during a reaction (cure); the symbol that this arrow follows is "given off"
−	A single bond between two atoms
=	A double bond between two atoms; quite reactive under certain conditions

IV. SUMMARY OF CURED ENCAPSULANT HARDNESSES

Flexible: from 0 to 80 A scale
Semiflexible: from 80 A scale to 50 D scale
Semirigid: from 50 D scale to 80 D scale
Rigid: greater than 80 D scale

V. IMPORTANT REACTIVE GROUPS

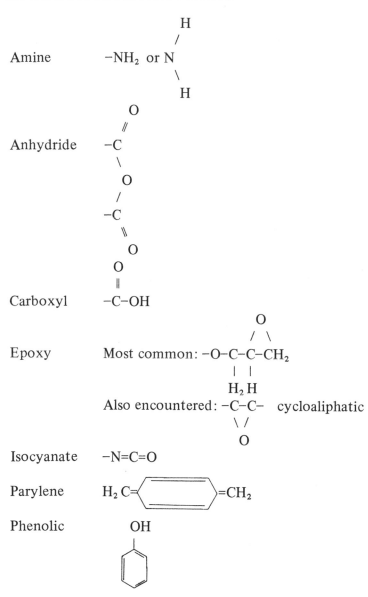

Amine

$$-NH_2 \text{ or } N \overset{\displaystyle \diagup \text{ H}}{\underset{\displaystyle \diagdown \text{ H}}{}}$$

Anhydride

$$
\begin{array}{c}
\quad O \\
\quad \| \\
-C \\
\quad \diagdown \\
\quad O \\
\quad \diagup \\
-C \\
\quad \diagdown\!\!\diagdown \\
\quad O
\end{array}
$$

Carboxyl

$$
\begin{array}{c}
O \\
\| \\
-C-OH
\end{array}
$$

Epoxy

$$
\text{Most common: } -O-C\overset{\displaystyle \overset{O}{\diagup \diagdown}}{\underset{\displaystyle \underset{H_2}{|}}{-}}C\overset{}{\underset{\displaystyle \underset{H}{|}}{-}}CH_2
$$

$$
\text{Also encountered: } -C\overset{}{\underset{\displaystyle \diagdown}{}}-C\overset{}{\underset{\displaystyle \diagup}{}}- \quad \text{cycloaliphatic}
$$
$$
\qquad\qquad\qquad O
$$

Isocyanate $-N=C=O$

Parylene $H_2C=\langle\!=\!=\!=\!\rangle=CH_2$

Phenolic   OH

Silicone

$$\begin{array}{c} CH_3 \\ | \\ -Si-OH \\ | \\ CH_3 \end{array}$$ condensation cure

$$\begin{array}{c} CH_3 \\ | \\ -Si-\left[\begin{array}{c} O \\ | \\ C \end{array}\right]H=CH_2 \\ | \\ CH_3 \end{array}$$ addition cure

Unsaturated $-C=C-$ polyester and DAP

VI. ELECTRICAL TEMPERATURE CLASSIFICATIONS

A material rated at a certain temperature is considered capable of continuous operation at that temperature.

Class (NEMA)	Temperature ($^\circ$C)	Class (military)	Temperature ($^\circ$C)
O	90	R	105
A	105	S	130
B	130	V	150
F	155	T	170
H	180	U	>170
E	215		

VII. VISCOSITY GUIDE

Viscosity [centipoise (cP)]	Common material
1	Water
10	Kerosene
50	SAE 10 oil
250	SAE 30 oil
650	SAE 50 oil

Viscosity [centipoise (cP)]	Common material
1,000	Glycerine
7,500	Liquid honey
10,000	Corn syrup
25,000	Chocolate syrup
100,000	Molasses

VIII. IMPORTANT ABBREVIATIONS

A. Epoxy Curing Agents

DETA	Diethylenetriamine
TETA	Triethylenetetramine
TEPA	Tetraethylenepentamine
NMA	Nadic methyl anhydride
HHPA	Hexahydrophthallic anhydride
MTHPA	Methyltetrahydrophthallic anhydride
DDSA	Dodecylsuccinnic anhydride
ATBN	Amine-terminated nitrile rubber
CTBN	Carboxyl-terminated nitrile rubber

B. Epoxy Types

Epi or Epi-Bis resins	Standard epoxy
EN	Epoxy novolac
ECN	Epoxy cresol novolac

C. Urethane Materials

TDI	Tolylene diisocyanate
MDI	Diphenylmethane diisocyanate
HMDI	Hexamethylene diisocyanate

D. Catalysts

BF_3 Boron trifluoride

DBT, T-12   Dibutyltin dilaurate
N-28   Stannous octoate
DMP10,
  DMP30*   Epoxy accelerator/catalysts
BDMA   Benzyl dimethylamine, epoxy catalyst/acceler-
                 ator

E. Molding Powders

DAP   Diallyl phthallate
BMC   Bulk molding compound

---

*Trademarks of Rohm & Haas.

2

Class I Materials

Class I materials are liquids that convert irreversibly to solids, with no significant weight loss, used as potting, casting, and encapsulating materials.

I. EPOXIES

A. General Formula and Description

$$\underset{CH_2-CH-CH_2-O-}{\overset{O}{\triangle}} \left\langle \bigcirc \right\rangle - \underset{CH_3}{\overset{CH_3}{\underset{|}{C}}} - \left\langle \bigcirc \right\rangle - O-CH_2-\underset{}{\overset{O}{CH-CH_3}}$$

Epoxies are the reaction product of two common organic chemicals, bisphenol A (Bis A) and epichlorohydrin (Epi), which when reacted in the ratio of 2 Epi to 1 Bis, a liquid reactive resin results—epoxy.

Epoxies are moderate- to low-viscosity nonvolatile liquids, without a characteristic odor and with a high flash point. They are low-to-moderate skin sensitizers. They react at room temperature with amine curing agents, and at higher temperatures with catalytic and anhydride curing agents to form a hard, infusible solid. Cured epoxies have medium to high heat deflection temperatures, but become quite brittle at temperatures below about 20°F.

Epoxies have a high dielectric strength and volume resistivity, and a relatively high dielectric constant and dissipation factor. Physically, epoxies display high tensile, compressive, and flexural strength, low impact strength, and very high adhesion to a wide variety of materials. They respond well to modifiers to adjust viscosity, hardness, and impact strength.

Virtually, all the epoxies used in class I are of the type pictured. The reactive points on the epoxy and the curing agents are matched 1:1, and this matchup must be maintained to achieve the proper cure, and development of the designed properties.

B. Advantages and Disadvantages

Advantages

Nontoxic (see Sections IC and ID)
High dielectric strength
High volume resistivity
Low viscosity
Inexpensive
High adhesion
Hard, can be modified

Disadvantages

High dielectric constant
High dissipation factor
Brittle at low temperatures

C. Curing Agents: An Overview

There are three *main* types of curing agents for epoxies:

1. *Amines.* Amines are nitrogen-based materials containing active hydrogen atoms that react with the epoxy group. A simple amine, ethylene diamine, is shown below. The hydrogens attached to the nitrogen are "active" and will react with an epoxy group.

   $H_2 N-CH_2-CH_2-NH_2$

2. *Acid anhydrides.* These are acidic-type chemicals containing anhydride groups which are "broken open" and react with the epoxy group. They require heat to cure. A simple anhydride, phthallic anhydride, is shown below.

3. *Catalysts.* Strictly speaking, catalysts are not curing agents. Catalysts cause the epoxy groups to react with each other and require heat to cure. Boron trifluoride (BF_3), a catalytic curing agent, is shown below.

   ```
   F   F
    \ /
     B
     |
     F
   ```

The reactive points of the amines and the anhydride groups must be matched with the epoxy groups on a 1:1 basis to achieve the desired and designed properties. This balance has been designed into the commercially available compounds. The designed properties will be obtained only if the recommended ratio of base to curing agent is maintained.

The catalytic types are added in very small amounts (1 to 5%) and cause the epoxy groups to react with each other. BF_3, boron trifluoride, is a common catalytic curing agent. It is never used as such, but always in a complexed form to stabilize it. Example: BF_3-MEA; here it is complexed with monoethyl amine. When heated, the complex breaks up, releasing the BF_3, which acts catalytically on the epoxy groups.

Part B, or a Choice. Many epoxy formulations come as "part A" and "part B" systems; that is, the curing agent is matched to the base (usually epoxy) component, and no substitution in type or amount is permitted. But many other epoxy materials are sold as a base material only, with a wide choice of curing agents offered, each giving a cured epoxy with different properties. Often, the chemical identity of these curing agents is known to the user, and sufficient data are supplied to permit a choice to be made. Because of this, the epoxy curing agents will be covered in considerable detail, to better equip the materials engineer to understand the capabilities and trade-offs available.

D. Curing Agent Details

1. Amines and Modifications

This class of chemical is used to cure more epoxies than all other curing agents combined, and exists in several forms. The general cure mechanism is shown below.

The amine group breaks the epoxy ring, allowing the nitrogen to connect to the epoxy. This leaves one hydrogen on the nitrogen available to react with another epoxy. Thus cross-linking is propagated as these multiended molecules react.

Amines are subdivided into two main classes: (1) aliphatic and their derivatives (which react at room temperature), and (2) aromatic, which require heat to cure.

Aliphatic Amines
General formula and description.

$$H_2 N[RN]_n \overset{\overset{\displaystyle H}{\displaystyle |}}{}H$$

where R is

$$[CH_2-CH_2]$$

and n is 2, 3 or 4.

Aliphatic amines are low-viscosity, almost colorless liquids. They exhibit a transient ammonia odor as the container is opened and a permanent, slightly "fishy" odor. They are used where a room-temperature curing system is necessary and where the operating temperature of the device will not exceed 130°C (class B), continuous.

Characteristics:

1. Low ratio (5 to 10%), difficult to measure accurately
2. Fast room-temperature cure
3. Develop considerable heat on cure
4. Can develop a surface blush or stickiness
5. Hard and often brittle
6. Can cause an allergenic response in certain people
7. Performance to class B (130°C)

Three commonly used amines are:

Diethylenetriamine (DETA): n = 2 in the general formula
Triethylenetetramine (TETA): n = 3 in the general formula
Tetraethylenepentamine (TEPA): n = 4 in the general
    formula

All are widely used, but usually coded as someone's catalyst or as curing agent "number so and so."

Points 3 and 6 will be explained more fully because of their potential for problems.

Point 3. The heat generated on cure restricts their use to small masses (usually less than 1/4 lb). As the mass of the mixed material increases, the heat cannot escape as fast as it is generated, so it builds up, sometimes to destructive levels. Either the device is damaged by the high temperature or the epoxy casting is distorted, with possible damage to the device it is there to protect. In extreme cases, internal charring can occur.

Point 6. The allergenic responses in some people can vary from a "poison ivy-like" skin rash, to eye and facial swelling, to respiratory difficulties. Aliphatic amines should be used with strong ventilation, with the air coming from behind the operator, over the workstation, then out through an exhaust system. Skin contact should be avoided, and plastic or rubber gloves should be standard practice, as should eye protection. If skin contact occurs, wash immediately with soap and water. If clothing is contaminated, remove it immediately and do not wear again until laundered *twice*.

An effective procedure is to have the operator apply a skin care lotion to the hands upon arrival at the beginning of each shift. If skin contamination occurs, remove with a waterless hand cleaner, using *no* water, wiping the hands off with paper towels, then reapply skin lotion. At shift end or lunch break, wash with soap and water, dry thoroughly, and reapply lotion. If eye contact occurs, flush with water and seek immediate medical attention.

At this point, the reader has probably written off aliphatic amines because of their potential problems. *Don't*!! These are precautions only. Aliphatic amines are used widely and successfully and do a fine job in their "niche."

Aliphatic Amine Adducts
General formula and description

$$\begin{matrix} H & & & H \\ \backslash & & & / \\ N-CH_2-CH_2-N & & + \text{ reactive material (R)} \rightarrow \\ / & & & \backslash \\ H & & & H \end{matrix}$$

$$\begin{matrix} \circ\!\!H & & & \circ\!\!H \\ \backslash & & & / \\ N-CH_2-CH_2-N & & \\ / & & & \backslash \\ \circ\!\!H & & & R \end{matrix}$$

where R can be an epoxy molecule or one of several other chemicals reactive toward amines. The circled "H's" are still reactive and will react with any other available epoxy groups.

Amine adducts are modifications of aliphatic amines to overcome some of the inherent disadvantages. The amine is prereacted with a small amount of epoxy or other chemical that is reactive toward the amine. There is still a large excess of unreacted amine left. This resultant "amine adduct" still reacts like the amine it is. However, the following improvements have been made:
Characteristics

Higher ratios of the adduct are required: 10 to 25%, thus permitting easier and more accurate measurements.
The adduct has a much lower vapor pressure, which means fewer and less severe allergenic responses.
There is less skin absorption in case of contact.
Viscosity is slightly higher.

The end properties of the cured epoxy are not changed appreciably in most cases, and the adduct can be used for the same end uses as the unmodified amine.

Aliphatic Amine Copolymer Modifications: Polyamides. In this group of amine modifications, the amine is used as one component in a further polymerization. Most often, products from vegetable oils are used as the coreactants with the amines, these being long-chain fatty acids which have been "dimerized"; that is, two monofunctional fatty acids have been joined together to form a dimer. When reacted with a multifunctional amine, a chain grows, which combines the hard brittle properties of the amines with the soft, flexible properties of fatty acids. These are polyamide curing agents.

General formula

$$\left[\begin{matrix} O && O \\ \| && \| \\ C\text{-}R\text{-}C\text{-}N\text{-} & R\text{-}N \\ & | & \quad H \\ & H \end{matrix} \right]_n H$$

where n has a value of approximately 5 to 15, and R, which represents the dimer acid and contains 34 carbon atoms, is very flexible.

Property changes versus unmodified amines

Much higher ratios for easier, more accurate measurements; even ratios of 1:1 or 2:1 by volume are possible.

Much lower vapor pressure (almost zero) and less skin absorption. Polyamides are the safest of the aliphatic amine-type curing agents to handle. Respiratory and skin problems in people handling polyamides is quite rare, provided that good housekeeping and good personal hygiene is practiced.

Higher viscosity.

Lower cured heat-deflection temperatures.

Less heat produced, meaning that larger castings can be made before temperature rise becomes troublesome.

Cure slower.

Amidoamines. There is one more class of modified amine which is in frequent use. It is known as an amidoamine, which, as the name implies, is somewhere halfway between an unmodified amine and a polyamide. Here, a fatty acid, this time *not* dimerized, is reacted with the amine, much as occurred in making an amine adduct. It is really a type of amine adduct.

General formula

$$
\begin{array}{c}
H \qquad\qquad H \\
\diagdown \qquad\qquad \diagup \\
N-CH_2-CH_2-N \quad O \\
\diagup \qquad\qquad \diagdown \;\; \diagup\!\!\diagup \\
H \qquad\qquad C \\
\qquad\qquad\qquad \diagdown \\
\qquad\qquad\qquad R
\end{array}
$$

where R is a 17-carbon-atom hydrocarbon chain.

Characteristics. As expected, amidoamine properties are partway between those of polyamides and amines.

Higher ratio than amines, but not as high as polyamides: 20 to 50%

Higher viscosity than amines, but much lower than polyamides

Much safer to handle than amines, but not quite as nice as polyamides

Slower room-temperature cure than amines, about the same speed as polyamides

Aliphatic Amine Property Summary

Unmodified amines	Amine adduct	Polyamide	Amidoamine
Lowest viscosity	Low viscosity	High viscosity	Medium to low viscosity
Small castings	Small castings	Large castings	Large castings
Hard brittle	Hard, less brittle	Not brittle, flexible	Not brittle, flexible
Fastest cure	Fast cure	Slow cure	Slow cure
Most difficult to handle	Difficult to handle	Easiest to handle	Easy to handle
Lowest ratio	Low ratio	Highest ratio	High ratio

Specialty Aliphatic Amines. There are some specialty aliphatic amine curing agents available, and although used rarely, the reader should be aware of them:

Flexible amines. These actually produce a flexible epoxy casting. They react very slowly at room temperature; a cure time of several days is normal. The cured castings show poor insulation resistance, which deteriorates further as the temperature rises.

Diethanolamine (DEA). This acts as both an amine and a catalyst. Long pot life. When completely cured (a higher-than-room-temperature post cure is recommended) will display good mechanical and thermal shock resistance.

Amine-terminated rubbers; amine-terminated butadiene acrylonitrile (ATBN). These have a dramatic effect on hardness and brittleness, decreasing hardness and greatly improving mechanical and thermal shock resistance. Very high viscosity and limited compatibility inhibit their widespread use in encapsulation epoxies. They find wider use in adhesives.

Aromatic Amines. Although a member of the amine family, aromatic amines differ markedly from aliphatic amines:

Differences from aliphatic amines

Require heat to cure
Very long pot life (hours)
Hard but *not* brittle castings
Service to 155°C (class F)
Very low allergenic potential

Aromatic amines offer significantly higher heat deflection temperatures and retention of electrical properties to higher temperatures than their aliphatic counterpart. They are also more difficult to remove chemically, offering improved tamper resistance. They will perform to class F standards (155°C). Although offering a higher ratio than aliphatic amines, the difference is not very great, so aromatics share the disadvantages of low ratios, with the attendant measurement-accuracy problems.

General formula and description

The mechanism of cure is the same as that for aliphatic amines. The one difference is that it occurs so slowly at room temperature that it is impractical, so heat, even as low as 50 to 60°C, is necessary.

Characteristics.   There is, however, one peculiarity in the cure of aromatic amines. At room or low temperature, a partial or B stage of cure is reached. At this point, the epoxy is hard but not cured. It is very brittle at room temperature and softens easily at temperatures over 125°F. It is stable in this stage for several days at room temperature, and actually is the basis for one type of epoxy powder used in fluid bed and compression molding (see Chapter 3). To complete the cure and toughening of the casting, more heat is required for a longer time.

Most aromatic amines cause stains when skin contact occurs, even briefly. The stain does not develop immediately, taking 1 to 2 days to show, and then appears as a brown to reddish-brown area. The only prevention is plastic or rubber gloves or complete noncontact. Recently, some modified versions have appeared that are claimed to be nonstaining.

Most aromatic amine curing agents will crystalyze upon standing. This is a physical phenomenon and is reversible with heat. These curing agents are a eutectic blend of two solid aromatic amines, metaphenylene diamine and methylene diani- line. When blended in the right ratio and melted, they become a liquid at room temperature. Crystallization occurs slowly and re- heating to 140 to 150°F is necessary to completely reliquefy. Vapors coming off during heating cause brown stains on every- thing with which they come in contact, including people, so heat- ing ovens should be vented to the outside.

A note of caution.   Since these two solid amines separately have different combining ratios to cure epoxies, it is important that partially crystallized amine not be used.   There is a risk that the liquid portion may be "rich" in one of the components and a "wrong" ratio could result.

2. Acid Anhydrides

Acid anhydrides (henceforth called "anhydrides") are a nonamine class of epoxy curing agents.

General Formula and Description

```
      H₂
      |
      C    H    O
     / |\ /      ‖
 H₂ C H C ——C
  |       |       \
  |       |        O
  |       |       /
 H₂ C   C_H —C
    \  /        ‖
     C          O
     |
     H₂
```

The anhydrides most commonly used are very low viscosity liquids, with high combining ratios, making measurements easier and fitting well into meter-mix-disperse equipment. They offer superior high temperatures (above 150°C), insulation resistance and dielectric strength, give a very long pot life (measured in days), and require high temperatures for several hours for a complete cure.

The cure mechanism of anhydrides is radically different from that of amines. The amine curing agent reacts on the basis of one amine hydrogen to one epoxy group. The anhydride group, however, causes the epoxy group to become bifunctional; that is, the epoxy molecule with a group at each end reacts as though there were two groups at each end. This results in a high cross-link density and improved high-temperature properties.

Advantages and Disadvantages
Advantages

High ratios, makes measuring easier
Low viscosity
High heat resistance
Superior electrical properties, particularly at elevated temper-
     atures

Disadvantages

High temperature for a long time required for cure
Require a "catalytic" accelerator to permit a cure in a reason-
     able time and temperature

Three Major Anhydrides.   The three most commonly used
anhydrides are (1) methyl nadic anhydride (MNA), (2) methyl
tetrahydrophthallic anhydride (MTHPA), and (3) hexahydrophthal-
lic anhydride (HHPA). HHPA is a solid at room temperature and
below, but melts easily and has no odor, unlike MNA and MTHPA.

Combinations with Anhydrides.   Liquid anhydrides are sometimes combined with dimer acids (which will also cure epoxies)—
seen previously in polyamides—to introduce some degree of flexibility into the cured system, while retaining good high-temperature
electricals. Heat-deflection temperatures would, of course, be
lowered.

Other Anhydrides.   There is some use of a series of the solid
aliphatic anhydrides: polyazelic polyanhydride (PAPA), poly-
sebacic polyanhydride (PSPA), and even phthallic anhydride (PA),
an aromatic material. They are all solids at room temperature and
are very inconvenient to use. They contribute very good mechanical and thermal shock properties; however, the difficulty in
handling them and their very long cure times discourage their use.
PA is probably the least expensive curing agent available. But its
use is discouraged because of the difficulty in melting it. It sub-

limes rather than melts, resulting in a white crystalline coating all over everything, and too little in the epoxy mix.

3. Catalytic Curing Agents

These agents are truly just that—*catalytic*. They cause the epoxy groups to react with each other and do not become a part of the cured system themselves as *do* the amine, anhydride, and fatty acid types. They require high temperature and several hours to effect complete cure, and are used in very small amounts (i.e., 1 to 5%).

General Formula and Description

$$BF_3 + nCH_2-\overset{\displaystyle O}{\overset{/\ \backslash}{CH-CH_2}}-R \rightarrow \ _n\boxed{\overset{\displaystyle \overset{|}{O}}{\underset{\underset{O}{|}}{\underset{|}{CH_2-CH-CH_2-R}}}}$$

Most catalytic curing agents are based on complexed (inactivated) forms of boron trifluoride (BF_3), a very reactive and corrosive chemical. Here the BF_3 is attached to another "protective" molecule which inactivates it. This complex is stable—unreactive—at room temperature. At temperatures above 100°C, it breaks up, releasing the BF_3, which then cures the epoxy through epoxy-to-epoxy group reactivity. These complexes enable one component (one package) epoxy systems to be formulated, which are stable for several weeks at room temperature.

BF_3 shares one characteristic with anhydrides when curing epoxies. That is, it also causes each epoxy group to react twice, and since the epoxy groups are reacting with each other, a very tightly cross-linked cured system results. Its characteristics include high heat deflection temperatures, rigidity, and brittleness.

Advantages and Disadvantages
Advantages

One-package systems are possible, thus eliminating mixing and measuring errors at the user level.

Hard, rigid, heat-resistant cured systems are produced.

Disadvantages

High temperatures are required to effect cure. Typical cure
times are 6 to 10 hr at 100°C, 2 to 4 hr at 120°C, and
½ to 1 hr at 140°C.

Very rigid and brittle cured systems may result.

Cannot heat a viscous one-component epoxy system to lower
the viscosity without adversely affecting the shelf
stability and pot life.

Certain pigments and fillers are not compatible with BF_3
complexes.

4. Other Heat-Cured Curing Agents

Dimer and Trimer Acids.   These curing agents are made by
connecting two (dimer) or three (trimer) long-chain fatty acids
together.

General formula and description

where R_1, R_2, and R_3 are 17-carbon-atom flexible hydrocarbon
chains.

One acid group reacts with one epoxy group. Because the
acid group is attached to a very long hydrocarbon chain, the resultant cured epoxy is characterized by very high flexibility. Unfortunately, the reaction takes place only at high temperatures,
225°F, and for long times, 12+ hr. Also, dimer and trimer acids
are not compatible with epoxies at room temperature, which

further restricts their use. When mixed with liquid epoxies, an opaque "lumpy pudding" liquid is the result. It becomes clear and free-flowing only at temperatures of 200°F and higher.

Advantages and disadvantages

Advantages

Good mechanical and thermal shock properties
Good retention of insulation resistance at elevated tempera-
    tures
High (variable) ratio (can adjust degree of flexibility)
Low cost

Disadvantages

Very slow cure
Very high temperature required to cure
Incompatible with epoxies at room temperature

E. Other Epoxies

Although the vast majority of epoxies used in protecting electrical/ electronic devices are depicted at the beginning of this chapter (known chemically as the Bis-A or Epi-Bis resins), other epoxy types are also used. In cure and use, these other types vary from simple modifications to radically different ones.

1. Cycloaliphatic Epoxies

General Formula and Description*

---

*Union Carbide ERL 4206.

Cycloaliphatics are not reactive toward amines, so room-temperature curing systems are not possible. They are cured with anhydrides and dimer acids. Their virtues are good electrical properties at high temperature and low viscosity. They are, however, very expensive, two to three times the cost of conventional Bis A resin, so their use is restricted to specialty applications. They also exhibit improved ultraviolet (UV)-light resistance, which permits their use in such outdoor applications as stand-off insulators and high-voltage bushings. With less surface breakdown (chalking), there is less dirt and dust contamination on the surface and less chance of a high-voltage flashover.

Advantages and Disadvantages
Advantages

Low viscosity
Good high-temperature electrical properties
Better UV-light resistance; used in stand-off insulators for
    outdoor use
Compatible at room temperature with dimer and trimer acids

Disadvantages

High-temperature cure required
Very expensive
No room-temperature cure

2. Rubber-Modified Epoxies

Specialty modified synthetic rubbers have been used to improve low-temperature flexibility and mechanical shock properties of the Bis epoxies. These are carboxyl-terminated butadiene acrylonitrile rubbers (CTBNs). They cannot be used as a stir-in additive since the carboxyl group reacts slowly with the epoxy group, resulting in a gradual viscosity rise. They should be "cooked" (i.e., reacted) into the epoxy portion. This uses up only a small portion of the available epoxy group, and the resultant modified epoxy is then cured by conventional means.

The CTBNs impart greatly improved fracture resistance to epoxy systems, but also result in very large increases in viscosity and slow cure. More recently, the previously mentioned ATBNs became available. These can be simply stirred into the curing agent portion, thus eliminating the cooking step. However, they suffer from limited compatibility as well as high viscosity and high cost, the latter shared by the CTBNs as well.

3. Epoxy Novolacs

There is a variant of the original Bis epoxy. Rather than epoxidizing bisphenol A, a short-chain phenolformaldehyde resin is prepared and subsequently epoxidized. This allows much higher (i.e., > 2) functionalities, which results in higher heat resistance and better chemical resistance.

General Formula and Description

where n = 0.5 to 2.0.

Epoxy novolacs are very high viscosity (approximately 50,000 cP at 50°C) liquid multifunctional resins. Unfortunately, this high viscosity precludes their use in normal liquid encapsulating systems, reducing their role to that of an additive, where increased heat and/or chemical resistance is desirable. Solid versions of these resins will be encountered in Section 1 of Chapter 3 as powdered molding compounds.

F. Fillers in Epoxies

Most epoxy compounds contain fillers. These are dry, powdery materials dispersed in the epoxy. They contribute many functional, as opposed to cosmetic, properties to the formulation:

Decrease exothermic heat buildup
Decrease shrinkage
Improve physical properties
Improve electrical properties
Impart "special" properties to the cured system (i.e., electrical conductivity, thermal resistance, etc.).

Quite frequently, the identity of the filler is reported in the material literature, so more filler detail will be given in Chapter 5.

II. SILICONES

Liquid RTV (room-temperature vulcanizing) silicone rubber compounds represent the second largest class of plastic materials used to protect electronic devices. Just like epoxies, when the correct materials are properly combined, they convert irreversibly to a solid with a very small or no weight loss.

A. General Formula and Description

$$
\begin{array}{cccc}
CH3 & CH3 & CH3 & CH3 \\
| & | & | & | \\
-Si-O-Si-O-Si-O-Si- \\
| & | & | & | \\
CH3 & CH3 & CH3 & CH3
\end{array}
$$

RTV silicones vary from low- to high-viscosity liquids, without a characteristic odor and with a high flashpoint. They are not primary skin sensitizers and present almost no handling hazards. They react with their curing agents (cross-linkers and catalysts) at room temperature to form very flexible thermoset solids. RTVs produce almost no heat during cure, and so, unlike many epoxies and urethanes, are no danger to themselves and what they protect.

Silicones undergo remarkably little physical change between their brittle point [i.e., -55°C (lower in some cases)] and their upper operating limit [i.e., +200°C (higher in some cases)]. They exhibit high dielectric strength and volume resistivity, and low dielectric constant and dissipation factor. Physically, silicones are quite weak, with low compressive, flexural, and tensile strengths but very high elongation. They differ radically from epoxies and urethanes in three characteristics:

> The coefficient of thermal expansion (CTE) is 5 to 10 times higher.
>
> They stick to nothing (except themselves); special primers permit "adequate" adhesion.
>
> They can be cut open with a scalpel very easily. This, combined with their poor adhesion, makes component replacement in potted units simple, fast, and easy.

B. Two Cure Mechanisms

RTV silicones have their reactive parts at each end of the molecule, just as epoxies do. However, there are two different groups used to cure silicones, and they are named for the type of cure reaction each type undergoes—addition and condensation cures—each with its own characteristics.

Note. Addition- and condensation-cure silicones *cannot* be mixed. At the very best, no cure will result.

1. Addition Cure

Description. In addition-cure silicones, the reactive parts simply add on to each other through a cross-linker, with nothing being eliminated, just as epoxies do. The reaction rate, *unlike epoxies*, proceeds so slowly that a catalyst is required to make the cure proceed at room temperature. The most commonly used catalysts are based on platinum, so addition-cure silicones are often referred to as platinum-cure silicones. The platinum catalyst can be added to either the base polymer or the cross-linker.

Advantages and Disadvantages
Advantages

Excellent deep-section cure at room temperature
Cure can be heat accelerated

Reversion resistant
High ratio of reactor to base component (10:1, even 1:1)

Disadvantages. The cure inhibits in contact with many materials commonly found in electronics. *Examples*: Amine-cured epoxies, many pressure-sensitive-tape adhesives, vulcanized rubber wire insulations, the "other" type of RTV silicones, some vinyl wire insulations, residual soldering flux, and even fingerprints can cause inhibition.

2. Condensation Cure

Description. With condensation-cure silicones, when the reactive parts connect with the cross-linker, a small amount of an alcoholic material is condensed out and eventually escapes. This is a very small weight percentage of the mass, but cure is not complete until this escapes. This reaction also proceeds very slowly, requiring a catalyst, in this case usually a tin-based compound. They are often referred to as "soap-cure silicones," referring to the composition of the tin catalyst.

Advantages and Disadvantages
Advantage. This is a noninhibiting cure.
Disadvantages

Slow deep-section cure
Very low catalyst ratios (0.1 to 0.5%), can be formulated to
     a 10% ratio
Can revert under certain conditions
Does not respond well to heat acceleration

C. Use Characteristics

RTV silicones produce almost no heat during cure and therefore are no danger to themselves (and what they protect), unlike many room-temperature curing epoxies and some urethanes.

```
CH3   CH3   CH3   CH3
 |     |     |     |
-Si-O-Si-O-Si-O-Si-
 |     |     |     |
CH3   CH3   CH3   CH3
```

Dimethyl silicone
Brittle point = -55°C

```
CH3   ⬡     CH3   CH3
 |     |     |     |
-Si-O-Si-O-Si-O-Si-
 |     |     |     |
CH3   CH3   CH3   CH3
```

Methyl phenyl silicone
Brittle point = -115°C

The general RTV silicone formula depicted is a standard "dimethyl" RTV. These silicones have a brittle point of approximately -50°C and a maximum operating temperature of 200°C. However, when some of the methyl groups attached to the Si atom in the backbone (the repeating unit) is replaced by a phenyl group, the brittle point is depressed to below -100°C, and the upper operating (continuous) limit is raised to about 250°C.

RTV silicones, as with epoxies, are available both as "matched sets" with the reactor portion mated to a corresponding base component, and as a base material, with a limited choice of catalysts. The term "catalyst" is used here as defined in Chapter 1.

Addition-cure silicone materials are always in a matched A/B set. The catalyst, usually a platinum compound, is used in such small amounts (in ppm) that it could not be handled accurately and cleanly by the user, so it is included in one of the components, either A or B. It can be added to the base silicone polymer or to the cross-linker. Cure takes place only when all three ingredients get mixed together.

Condensation-cure silicones, on the other hand, permit a small choice of catalysts by the user. They are all tin compounds: one, a regular (overnight) cure (dibutyltin dilaurate, commonly referred to as T-12*) and the other, a fast cure, 1 to 2 hours (stannous octoate, commonly referred to an N-28†). They are used in amounts of 0.1 to 0.5%, which permits either weighing (for large amounts) or counting drops (for small amounts). Both tin compounds are available in formulated versions that permit the amount added to be raised from 0.1–0.5% to 10%, to be measured much more easily and accurately, and are better suited to handling in meter-mix-dispense equipment.

D. Advantages and Disadvantages

Advantages

Easily cut open for component repair or replacement
Relatively constant electrical and physical properties over a
very wide temperature range, from their brittle point to
the upper operating temperature limit
Widest operating temperature range of all the encapsulating
materials
Highest insulation resistance and dielectric strength
Low dielectric constant and dissipation factor
No exotherm during cure
No handling hazards, although the tin soaps can be an irritant

Disadvantages

Very expensive compared to epoxies and urethanes
Low physical strength
No adhesion (fair with special primers or with self-stick no-
primer-required RTVs, available beginning late in 1984)
Very high coefficient of thermal expansion
Addition-cure types are easily inhibited from curing
Condensation silicones cure slowly in deep sections

---

*Trademark of M & T Chemical Co.
†Trademark of Nopco Chemical Co.

E. Handling

RTV silicones are handled just like epoxies; the right amount of catalyst (curing agent in the case of the epoxies) is mixed into the base material. The working life is, of course, dependent on the catalyst and amount used, and the ambient temperature. But unlike epoxies, silicones do not evolve heat during cure, and thus do not suddenly get hot, and abruptly gel toward the end of their working life. Silicones progressively thicken until they are no longer flowable. After 4 to 6 hr, the device can usually be handled, and is fully cured overnight.

F. Fillers in Silicones

Mineral fillers dispersed in the silicone polymers provide the same functional properties of thermal conductivity, and so on, that they do in epoxies. However, they are essential in most silicone formulas, as they provide the basic physical strength inherent in epoxies and urethanes without fillers but which is lacking in cured silicone polymers. The one exception is an unfilled, clear, strong silicone gel, which is modified for better physical strength another way. We have more to say about fillers in Chapter 5.

III. URETHANES

A. General Formula and Description

$$O=C=N-R-N=C=O$$

A diisocyanate

Isocyanates are the basis of polyurethane encapsulants. They are analogous to the epoxy portion of epoxy encapsulants and as such, must be cured with appropriate hardeners. Like epoxies, they have their reactive areas at each end of the molecule.

Urethanes are unusual in that the reactive isocyanate group breaks up and rearranges upon cure, becoming quite unrecognizable.

Polyurethanes are named for the reaction product that results from the interreaction of an isocyanate and a hydroxyl group. To be chemically correct, the reaction product of an isocyanate and an amine should be called a substituted urea.

Polyurethane compounds generally are low-viscosity, two-component liquids, which in appearance are not unlike many epoxies. They have little odor, a moderate to high flashpoint, and react rapidly at room temperature with a modest evolution of heat (less than epoxies, more than silicones) to a cross-linked solid plastic, varying in hardness and flexibility from semirigid to flexible.

There is one characteristic in which urethanes are very different from epoxies and silicones. They are very moisture sensitive. The isocyanate component must be protected from moisture, including humidity. The isocyanate group will preferentially react with any available moisture to which it is exposed, resulting in decreased reactivity toward its curing agent and sometime foaming during cure. Suppliers usually recommend that opened cans be backflushed with dry nitrogen before reclosing.

Cured urethane systems (polyurethanes) have good low-temperature flexibility, poor high-temperature properties, good dielectric strength and insulation resistance, and medium to low dielectric constant and dissipation factor. They have a very short pot life and a moderate exotherm, which can, however, become troublesome in larger masses.

All urethane systems come as matched A/B sets, so the user has no freedom to vary the properties by selecting a different curing agent (part B), as is done frequently with epoxies and to some extent with silicones. For this reason, no time is expended on exploring the effect of various types of amines, glycols, and so on, on the end urethane properties. One very strong reason for this lack of freedom of choice is the extreme moisture sensitivity of the isocyanate reactive group. Even if a cross-linker (an amine, for example) could be obtained that would yield a more satisfactory cured urethane than the curing agent provided, and even if adequate data were provided to permit the new cross-linker to be used in the correct amount, failure would probably result. *Reason*: All ingredients in a urethane formulation must be urethane grade (very dry). When a "wet" ingredient is used, two competing reactions occur: (1) the desired "cure" reaction, and (2) the undesired water-isocyanate reaction, which causes evolution of a gas (carbon dioxide, CO_2) and foaming.

$$\text{Desirable reaction} \rightarrow \underset{\underset{H}{|}}{R}\overset{\overset{H}{|}}{N}-\overset{\overset{O}{||}}{C}-OR$$

$$\text{Undesirable reaction} \rightarrow \underset{\underset{H}{|}}{R}\overset{\overset{H}{|}}{-N} + CO_2 \uparrow$$

When a gas is evolved in an encapsulant, obviously the application is a failure and the device destroyed, or at the very least, in jeopardy.

B. Isocyanate Sources

There are several isocyanate-containing materials available to the formulator, but mainly two are used: (1) tolylene diisocyanate (TDI), and (2) diphenylmethane diisocyanate (MDI). Both are extremely moisture sensitive and somewhat volatile and relatively toxic. For these reasons, they never exist "as such" in urethane systems, but as a prepolymer, where the diisocyanate has been pre-reacted with part of the cure system, leaving an excess of isocyanate groups to permit it to react with the balance of the cure system. This serves to lessen the moisture sensitivity to make the isocyanate less volatile and to raise the ratio of isocyanate to reactor in the end formulation. This is the same technique as that used with low-ratio volatile amines used as epoxy curing agents, when they are adducted. These, too, could be correctly referred to as prepolymers. This prepolymer is then matched to the balance of the reactor ingredients and packaged as a matched pair—part A and part B—with a fixed ratio of one to the other.

C. Urethane Cross-linkers

The list of chemicals that will react with an isocyanate group is a long one. It includes alcohols, glycols, water, acids, amines, and so on. Even natural sugars are used as cross-linkers in making some urethane foam products. However, in electronic encapsulation the cross-linkers usually come from two main classes:

1. *Amines*: such as the epoxy curing agents
2. *Hydroxyls*: usual sources are the polyethers (polyglycols) and hydroxyl-terminated polyester resins

The amine types give harder, more rigid cures; the hydroxyl-containing materials give softer, tougher, more flexible cures. The soft semiflexible urethanes offer low stress on components, but are still too tough to be cut and picked apart easily for component repairs and replacement, as is done with silicones.

D. Advantages and Disadvantages

Advantages

Fast cure
Moderate exotherm
Good low-temperature flexibility
Low viscosity
High ratios of A to B
Moderate cost, in the range of epoxies
Excellent thermal and mechanical shock resistance
Low dielectric constant and dissipation factor

Disadvantages

Very moisture sensitive
Poor high-temperature properties: electrical and physical
Many will "revert" (see the definition of "reversion" in
    Chapter 1)

Short working life (pot life)
High coefficient of thermal expansion

E. Use Characteristics

Polyurethane compounds are handled much like epoxies and silicones *except* that account must be taken of their moisture sensitivity and appropriate precautions exercised. Also, since pot life is
shorter, application to the device must take place more quickly
than with epoxies and silicones. Just as with epoxies, the prescribed amounts of the two components are mixed *thoroughly* and
quickly. The gel point is usually very sharp, converting from a
liquid to a solid quite abruptly, with a modest exothermic temperature rise. The resultant cured material is tough and resilient, dis-
playing excellent mechanical and thermal shock resistance, but a
low upper operating temperature. Urethanes find use in a broad
area of electronic protection where performance to class B is required.

F. Fillers in Urethanes

Urethane compounds are not usually filled with mineral fillers,
unlike most epoxies and silicones. The principal reason is the
need, and difficulty, in getting the moisture content of the fillers
low enough to prevent the previously described and very undesirable reaction with the isocyanate group, with resulting
foaming due to CO_2 production. Thus, with lower filler levels,
urethanes tend to shrink more on cure than do epoxies and
silicones. There are a few filled urethane formulations available,
so it is certainly possible, just difficult. Where certain physical
or electrical properties available only by use of specific fillers
are essential, the fillers must be dried out prior to their addition to
the formulation.

IV. UNSATURATED POLYESTERS

A. General Formula and Description

$$R\left[O-\overset{O}{\overset{\|}{C}}-C=C-\overset{O}{\overset{\|}{C}}-O\right]_n R \ + \ \bigcirc\!\!-\!\!\overset{C=CH_2}{\underset{H}{}}$$

General polyester formula   Styrene

Styrene is an ever-present and necessary reactive diluent in polyesters, and accounts for their characteristic odor.

The quantity of unsaturated polyester resin used in electronic encapsulation has diminished almost to the vanishing point. Very high shrinkage, high exotherm, high odor, flammability in the wet state, and poor adhesion have offset the low viscosity and attractive lower price (about two-thirds that of epoxies).

Polyesters, like RTV silicones and unlike epoxies and urethanes, cure by catalytic action. Unfortunately, the reaction is inhibited easily (even by air), so precautions must be taken to avoid a sticky, uncured surface or larger areas of uncured material in the interior. There is no cross-linker involved in curing polyesters. The reactive points are spaced along the backbone chain rather than at the end as we have seen in the previous materials. Under the influence of the catalyst, the reactive points react directly with each other and with the styrene diluent.

Polyesters use small amounts (0.1 to 1%) of a catalyst, usually MEKP (methyl ethyl ketone peroxide) to effect cure. Various accelerators and promoters are used to accelerate and promote the reaction, so a very short pot life and very fast cure are normal, accompanied by a very high exotherm. This exotherm can result in an even-higher-than-expected shrinkage. Normal polyester shrinkage is 3 to 5%, even if the high exotherm is handled.

B. Advantages and Disadvantages

Advantages

Low cost
Fast cure
Low dielectric constant and dissipation factor
Lower viscosity

Disadvantages

Very high shrinkage
Cure can be inhibited easily
Very flammable in wet, uncured state
Flashpoint $< 100°F$
High odor
Poor shelf life (uncatalyzed), should be stored in cool environment
Low maximum operating temperature $\simeq 100°C$ or less, depending on formulation
Very high exotherm
Solvent sensitive

C. Use Characteristics

Polyester compounds are handled in use, much like condensation-cure silicones, in that a small amount of catalyst is added and the reaction starts. Because so much heat is generated, they cure much faster than silicones, however, so they must be applied to the device almost immediately. The conversion from liquid to solid occurs quickly, and the gel point is very abrupt. Unless controlled, the heat produced can be very destructive, so cooling may be required. The resultant cured material is solvent sensitive with poor adhesion. Flames must be excluded from the handling area, and good ventilation is essential.

   As with urethanes and silicones, the user has no freedom to adjust the end properties by selection of a different curing agent.

This must be done by altering the "backbone" of the polyester itself or by a formulator adjusting the ingredients. Even this is difficult to impossible, since polyesters do not lend themselves to the "formulate, store, ship, and use" methods of the preceding three polymers, because of poor storage stability and extreme sensitivity to storage temperatures. Normally, a user must buy the raw polyester, styrene diluent, and MEKP catalyst, then become an "on-site" formulator. Polyester encapsulation systems should not be used at temperatures above 175 to 200°F.

D. Fillers in Polyesters

The filler most used in polyesters is sand, usually not premixed in. Rather, the sand is put into the device prior to application of the polyester. The resin is then poured over the sand-filled unit and percolates down and through the sand, subsequently curing.

V. OTHER LIQUID ENCAPSULATING POLYMERS

There are a few other liquid polymer types that have been used in electrical/electronic encapsulating compounds. They are infrequently encountered and are presented here only to give the reader a brief description of their properties.

A. Polybutadiene

Butadiene, the monomer, is a major raw material in synthetic rubber manufacture, particularly in "synthetic natural" rubbers. If polymerized with one catalyst system, an elastomer—soft, flexible, and extensible—results. If polymerized with a different catalyst, a resinous material results. Both types are used in electrical/electronic applications.

1. Elastomeric Polybutadiene

Description. This is a low-viscosity liquid version of the heavy gum rubber used to mold rubber goods and tires. When cured through a catalytic mechanism, the polybutadiene becomes a tough, flexible, highly extensible rubber in the A hardness range 25 to 50. It exhibits low shrinkage and very low exotherm on cure. Its insulation resistance is good at moderate temperatures, but it drops off rapidly at temperatures over 100°C. It should be classified as a low-temperature-cure system rather than a room-temperature cure. If it is to be used above 100°C, a postcure of 125 to 150°F is required to complete full property development.

Elastomeric polybutadiene, being a pure hydrocarbon, is quite flammable unless specially formulated to become self-extinguishing. It is somewhat easier to repair than soft epoxies and urethanes but still not like an RTV silicone. It is both reversion resistant and hydrolytically stable when fully cured.

Advantages and Disadvantages
Advantages

Cured rubber very soft for low stress
Relatively low viscosity
Fast gel at room temperature
Low exotherm on cure
Low shrinkage on cure
More reversion resistant than urethanes or condensation-
    cure silicones

Disadvantages

Not recommended for service over 130°C
Flammable
Elevated-temperature cure necessary for service higher than
    150°C
Relatively more expensive than epoxies and urethanes

2. Resinous Polybutadiene

Description. When polymerized the "resinous" way, buta-diene yields a low-viscosity liquid, which when cured yields a hard, infusible solid in the D hardness range 60 to 90. These polybuta-dienes are usually formulated into a one-component, heat-cured impregnating material for high-temperature (to 180°C) use in coils and transformers. Although these materials can be "mass" cured, the high cure temperature required and the resulting shrinkage makes them best suited for use as 100% solid impregnating varnishes. Cure times and temperatures are in the range from 4 hr at 150°C to 1 to 2 hr at 180°C. These polybutadienes, unlike their rubbery brothers, give excellent high-temperature electricals, even to 180°C.

Advantages and Disadvantages
Advantages

Excellent high-temperature electrical
Low viscosity
One-component system
Excellent reversion resistance and hydrolytic stability

Disadvantages

Very high temperature cure required
Generally unsuitable for mass casting
Relatively high shrinkage
Some compatibility problems when cured in contact with certain materials containing phenolic resins, sulfur compounds, rubber, and so on.

B. Depolymerized Rubber

1. Description

There is a process by which natural and synthetic rubbers, which exist as high-viscosity liquids or "gum" solids (before cure), can be depolymerized back down to lower-viscosity liquids. They can now be cured to soft elastomeric rubbers. They have found some limited use in low-cost, low-stress potting systems when price is of paramount importance. The electrical properties are poor above 150°F. They cure with no exotherm and very low shrinkage. The viscosity of the various DPRs is generally higher than that of the polybutadienes, and they are cured using metal salts, among which

are lead compounds. Natural, synthetic natural, and butyl rubbers
are all used in making DPRs.

2. Advantages and Disadvantages

   Advantages

      Very soft, low stress
      Low cost
      Low shrinkage
      Low exotherm

   Disadvantages

      Low tear strength
      Poor electricals over 150°F
      Some contain lead compounds
      High viscosity

C. Polysulfide Polymers

1. Description

As the name suggests, these materials contain sulfur in the backbone polymer chain. As well, a hydrogen-capped sulfur group
forms the reactive mechanism at each end of the molecule. It is
through this group that the cure takes place. Polysulfides are relatively low viscosity liquids with a pronounced odor, often on the
"skunky" side. Their use in electronics/aerospace is mainly in the
form of sealants with good resistance to ozone, fuels, and chemicals. They can and occasionally are cured with liquid epoxies together with amines as part of the cure. They improve epoxies' impact resistance and low-temperature flexibility, but their use has
dropped off in the past few years. Their high viscosity, pronounced odor, and relatively high cost (approximately twice that
of liquid epoxies or urethanes) diminishes their use to special applications.

APPENDIX: SOURCES FOR MATERIALS

Liquid Epoxy Resins

Celanese, Louisville, Kentucky
Ciba-Geigy, Ardsley, New York
Dow Chemical, Midland, Michigan
K & S Industries, Weston, Connecticut
Reichhold Chemical Co., North White Plains, New York
Shell Chemical, Houston, Texas

Epoxy Curing Agents

Aliphatic Amines

Dow Chemical, Midland, Michigan
Union Carbide, Danbury, Connecticut

Modified Aliphatic Amines

A.Z.S. Co., Eaton Park, Florida
Celanese, Louisville, Kentucky
Ciba-Geigy, Ardsley, New York
D.H. Litter, New York, New York
Jefferson Chemical, Houston, Texas
K & S Industries, Weston, Connecticut
Pacific Anchor Chemical Co., Richmond, California
Reichhold Chemical Co., North White Plains, New York
Shell Chemical, Houston, Texas
Trimont Chemicals, Peabody, Massachusetts

Polyamides/Amidoamines

A.Z.S. Co., Eaton Park, Florida
Celanese, Louisville, Kentucky
Ciba-Geigy, Ardsley, New York
D.H. Litter, New York, New York

General Mills, Minneapolis, Minnesota
K & S Industries, Weston, Connecticut
Pacific Anchor Chemical Co., Richmond, California
Reichhold Chemical Co., North White Plains, New York
Shell Chemical, Houston, Texas
Trimont Chemicals, Peabody, Massachusetts

Acid Anhydrides

Anhydrides & Chemicals, Newark, New Jersey
Celanese, Louisville, Kentucky
Ciba-Geigy, Ardsley, New York
D.H. Litter, New York, New York
K & S Industries, Weston, Connecticut
Lindride Chemical,
Pacific Anchor Chemical Co., Richmond, California
Shell Chemical, Houston, Texas

Epoxy Catalysts

Argus Chemicals, Brooklyn, New York
Ciba-Geigy, Ardsley, New York
Pacific Anchor Chemical Co., Richmond, California
Rohm & Haas, Philadelphia, Pennsylvania
Sherwin-Williams, Chicago, Illinois

Dimer/Trimer Acids

Emcry Industries, Cincinnati, Ohio
Humko Chemical Co., Memphis, Tennessee
Union Comp., Jacksonville, Florida
Witco Chemical Co., Detroit, Michigan

Cycloaliphatic Epoxies

Ciba-Geigy, Ardsley, New York
Union Carbide, Danbury, Connecticut

Epoxy Novalocs

Ciba-Geigy, Ardsley, New York
Dow Chemical, Midland, Michigan
K & S Industries, Weston, Connecticut

Isocyanates

Allied Chemical, Morristown, New Jersey
Dow Chemical, Midland, Michigan
Mobay Chemical, Pittsburgh, Pennsylvania
Olin Corp., Stamford, Connecticut
Wilmington Chemical, Wilmington, Delaware

Hydroxyl Sources

Allied Chemical, Morristown, New Jersey
Ashland Chemical, Columbus, Ohio
Atlas Chemical,
Celanese, Louisville, Kentucky
Dow Chemical, Midland, Michigan
Eastman Chemical, Kingsport, Tennessee
Hercules Chemical Co., Wilmington, Delaware
Jefferson Chemical, Houston, Texas
Mobay Chemical, Pittsburgh, Pennsylvania
Olin Corp., Stamford, Connecticut
Shell Chemical, Houston, Texas
Spencer Kellogg, Buffalo, New York
Stouffer Chemical, Westport, Connecticut
Union Carbide, Danbury, Connecticut
Witco Chemical Co., Detroit, Michigan
Wyandotte Chemical, Wyandotte, Michigan

Isocyanate Amine Sources

Allied Chemical, Morristown, New Jersey
American Hoescht, New York, New York

Celanese, Louisville, Kentucky
Ciba-Geigy, Ardsley, New York
Dow Chemical, Midland, Michigan
Du Pont de Nemours, Wilmington, Delaware
Jefferson Chemical, Houston, Texas
Shell Chemical, Houston, Texas
Union Carbide, Danbury, Connecticut
Uniroyal, Southfield, Connecticut
Wyandotte Chemical, Wyandotte, Michigan

Liquid Polyester Resins

Allied Chemical, Morristown, New Jersey
American Cyanamid, Wayne, New Jersey
A.Z.S. Co., Eaton Park, Florida
Cargill Co., Minneapolis, Minnesota
Reichhold Chemical Co., North White Plains, New York
U.S.S. Chemical, Pittsburgh, Pennsylvania

Silicone Catalysts

Cardinal Chemical Co., Columbia, South Carolina
M & T Chemicals, Rahway, New Jersey

3
Class II and III Materials

I. CLASS II MATERIALS

The end use of class II materials finds them usually classified as "molding materials," meaning that they will be heated and pressed—molded—into a shape. They are all thermoset materials as opposed to the better known and much larger volume thermoplastic molding materials.

Thermoplastics are softened by heat, shaped, then cooled to the solid state again, this time in the shape of the mold. This process can be repeated, since no chemical reaction or "cure" took place.

Thermosets are also softened by heat, shaped, then cooled. However, they actually chemically react or "cure." This cure is irreversible and they cannot be heat softened and reshaped.

Class II materials are received by the user as solids, usually finely divided powders, or as compacted slugs called "preforms." These powders are heated and become liquid for a short period, during which they are applied, then they cure to the final irreversible thermoset condition. This cure nearly always occurs in a metal mold.

With class I materials, heat is incidental to application, being used sometimes to lower viscosity, whereas heat is essential to the application, as well as the cure of class II materials.

These materials of class II are blends of solid reactive plastics, such as solid epoxies, phenol-formaldehyde (phenolic) resins, and diallyl phthalate (DAP) resins, with appropriate mineral fillers, including reinforcing fibers and cross-linkers, catalysts, release agents, and so on. They are always one-component and may have to be shipped and stored under refrigeration to prolong their useful life.

There is a variant member of class II materials, polyester bulk molding compound (BMC). It is made from the type of liquid polyester resin we saw in class I materials, but formulated with a high level of glass fibers, dry mineral fillers, and appropriate catalysts and accelerators. The resultant mixture is a dough consistency and is usually shaped into an extruded log. Although not strictly a powder, it is quite dry to the touch and is used in many of the same end uses as the powdered members of class II.

When class II materials are heated, two things happen simultaneously: (1) the powder (or BMC) liquefies and under pressure is forced into a cavity, and (2) the powder begins to cure. Now we have a race, always won by the cross-linking mechanism, but hopefully, arranged so that the material has reached its destination (i.e., filling the mold) before cure is effected, and the infusible mass cannot be shaped or moved further.

A. Diallyl Phthalate

1. General Formula and Description

$$
\begin{array}{l}
\text{C} - \text{O} - \overset{\text{H}_2}{\text{C}} - \overset{\text{H}}{\text{C}} = \text{C} - \text{H}_2 \\
\text{C} - \text{O} - \text{C} - \text{C} = \text{C} - \text{H}_2
\end{array}
\quad + \text{ catalyst } \xrightarrow{\ 325^\circ F\ } \text{thermoset DAP}
$$

2. Advantages and Disadvantages

Advantages

Very fast cure allows short cycle times
Hard and tough
Long shelf life unrefrigerated

Disadvantages

Exert high pressure during molding, therefore difficult to mold around inserts
Expensive
High-capital-cost equipment

3. Use Characteristics

When DAP powder is heated and pressured, it melts, the catalyst is activated, and the now molten mass flows, then quickly cures. The cycle is complete in a minute or two. DAP is used to mold small parts, switches, connectors, terminal blocks, or even potting shells for devices potted in class I liquid materials. Molding temperatures are 300 to 325°F. The cured plastic is extremely hard, rigid, tough, and heat resistant.

B. Epoxies

1. General Formula and Description

$$H_2-\overset{\overset{\displaystyle O}{}}{\underset{\underset{\displaystyle H}{}}{C}}-\overset{}{C}-CH_2 \left[O-\bigcirc-\overset{\overset{\displaystyle CH_3}{}}{\underset{\underset{\displaystyle CH_3}{}}{C}}-\bigcirc-O-\overset{\overset{\displaystyle H}{}}{\underset{\underset{\displaystyle H}{}}{C}}-\overset{\overset{\displaystyle O}{}}{\underset{\underset{\displaystyle H}{}}{C}}-\overset{\overset{\displaystyle H}{}}{\underset{\underset{\displaystyle H}{}}{C}} \right]_n -O-\bigcirc-\overset{\overset{\displaystyle CH_3}{}}{\underset{\underset{\displaystyle CH_3}{}}{C}}-\bigcirc-O-\overset{\overset{\displaystyle H}{}}{\underset{\underset{\displaystyle H}{}}{C}}-\overset{\overset{\displaystyle O}{}}{\underset{\underset{\displaystyle H}{}}{C}}-CH_3 +$$

$$\text{cross-linker} + \text{catalyst} \xrightarrow{\,\simeq 325^\circ F\,} \text{thermoset epoxy}$$

thermoset epoxy

Epoxy powders are a solid version of the liquid epoxy of class I. The chemistry of cure is the same, but it is in a dry, solid form. The formulated powder contains a latent curing agent, an accelerator, mineral fillers, sometimes short glass fibers, and a mold release.

2. Advantages and Disadvantages

Advantages

Very high flow
Very fast cure
Hard
Excellent electrical properties
Many small components can be quickly encapsulated in one
    step by use of a multicavity mold

Disadvantages

Expensive, more than DAP
Short room-temperature shelf life, 90 to 120 days
Best kept refrigerated
Capital-intensive equipment required

3. Use Characteristics

The action of an epoxy powder under heat is the same as that of a DAP. The powder melts, the curing agent and epoxy begin to react, and the reaction is accelerated by the now molten accelerator. Just as with DAP, the liquid charge is forced into the shape of the mold cavity, and then cures. Epoxies are generally used to encapsulate coils as well as small components such as integrated circuits and resistors, in high volume. Molding temperatures are the same as for DAPs. Thin walls tend to be more brittle than DAPs and phenolics.

C. Phenolformadehydes and Phenolics

1. General Formula and Description

Note.   Ammonia is given off. This usually necessitates an additional step, that of opening and reclosing the mold during the cure cycle, to vent the ammonia vapors.

Phenolic powders in composition are rather like epoxies in that they contain the basic phenolic resin, a cross-linker (curing agent), an accelerating catalyst, mineral fillers, short fibers, and a mold release. They cure to a very hard, tough solid with good chemical and solvent resistance, are very hard to strip chemically, and have good electrical properties and high heat resistance.

3. Advantages and Disadvantages

Advantages

Resistant to chemical removal
Cheaper than epoxies or DAPs
Very long room-temperature shelf life

Disadvantages

Higher shrinkage on cure than epoxies or DAPs
Higher cure temperatures, 325 to 400°F

3. Use Characteristics

Phenolics mold much like epoxies, but somewhat more slowly. Because of their high strength and heat resistance, phenolics are most often used to mold parts: bobbins and terminal blocks as well as potting shells. The ammonia given off during cure is the cause of greater shrinkage than with epoxies and often necessitates opening the mold to vent during the cure cycle.

D. Bulk Molding Compounds

1. General Formula and Description

$$\left[R-O-\overset{\overset{\displaystyle O}{\|}}{C}-C=C-\overset{\overset{\displaystyle O}{\|}}{C}-O\right]_n R- \;+\; \text{catalyst} \xrightarrow{\;\simeq 340°F\;} \text{thermoset polyester}$$

Bulk molding compounds (BMCs) are the anomoly of the series because they are not dry powders but rather, of a doughy consistency. The basic liquid polyester resin, like that used in the class I materials, is mixed with fillers, fibers, accelerators, and a latent peroxide catalyst, to a dough consistency.

Cured BMC is very tough, hard, and impact resistant, but with a lower softening and service temperature than those of epoxies, DAPs, and phenolics. Its low-temperature (below 200°F) properties are good, but solvent and chemical resistance are poor. They are, however, significantly cheaper than any of the foregoing.

2. Advantages and Disadvantages

Advantages

Very fast cycle times
Inexpensive

Disadvantages

Very short shelf life (about 60 days), so should be refrigerated
Low operating temperature

Strong odor in wet state

More flammable in wet state

3. Use Characteristics

Application of heat has the same effect as with the other three materials: soften, flow, and cure. BMC is normally shipped as a preformed log. A predetermined amount is cut off and placed in a cavity. BMC molds faster than the other materials in this group and is used in plates, bases, separators, and potting shells.

E. Solid Epoxy Novolacs and Solid Epoxy Cresol Novolacs

1. General Formula and Description

Epoxy cresol novolac   Novolac epoxy

These two solid epoxy-type resins are similar, varying only by the addition of a methyl group on the ring of the cresol types. They react similarly to the liquid epoxy novolacs encountered in Chapter 2. They are higher-molecular-weight homologs, which are solid, similar to the liquid Bis epoxies, being extended in molecular weight until they are solid. This facilitates their use in molding powders. Although they will respond to the same curing agents as their liquid cousins, a powder curing agent is more useful in a powder compound. Frequently, ENs and ECNs are cured with a phenolic or cresol resin similar to their starting resins, but *not* epoxidized.

Look at the structural formulas and imagine the epoxy groups removed and replaced by a hydroxyl, OH. Normally, this phenolic hydroxyl is not very reactive toward the epoxy group, but the increase of reactivity of the EN and ECN structure, the use of high levels of catalytic-like accelerators, and the very high temperatures at which they are moded cause this reaction to proceed.

In composition, the formulated powders are much like their Bis epoxy counterparts, containing fillers, reinforcements, release agents, and colorants. Where even better moisture/water resistance is required, a silane coupling agent will be used to improve the epoxy/filler interface bond. They gel quickly and cure to a very hard, very heat resistant solid with superior electrical properties, particularly at temperatures above 150°C. The ENs have an edge on heat deflection temperature over the ECNs but the later are said to mold a little more easily.

2.   Advantages and Disadvantages

Advantages

Very good high-temperature properties
Very low hydrolyzable chloride levels, important in semi-
    conductor molding
Very fast gel times
Excellent electrcal properties at temperatures over 300°F

Disadvantages

Much more expensive than the other solid materials of class
    II
Very high molding temperatures with very high postcure re-
    quired for maximum property development

3.   Use Characteristics

EN and ECN powders are handled just like their Bis epoxy counterparts, but are molded at higher temperatures. They are used to mold semiconductors and devices requiring very high heat resistance and strength and/or chemical resistance at high temperatures. Their chemical resistance is much superior to Bis epoxies and DAPs. The novolacs are also used in blends with solid Bis resins to shorten their gel times and upgrade their heat resistance and strength.

F. Summary of Class II Materials

Property	DAP	Epoxy	Phenolic	BMC	EN	ENC
Cost	High	High	2nd	Lowest	Highest	Highest
Operating temperature	Higher	High	High	Low	Highest	Highest
Shelf life	Long	Short	Longest	Shortest	Long	Long
Toughness	Good	Poorest	Very Good	Good	Good	Good
Cycle time	Second	Third	Slowest	Fastest	Fastest	Fastest
Chemical/solvent resistance	Medium	Medium	Very Good	Poorest	Excellent	Excellent

II. CLASS III MATERIALS

The class III materials are liquids as received. They usually consist of a viscous liquid or solid material, dissolved in a solvent, from which the solvent evaporates once the material is applied, leaving a thin film of a solid material. Depending on the type of starting material (i.e., acrylic, epoxy, urethane, etc.), the evaporation of the solvent may be accompanied by a "curing" reaction as well. The solvent is a "fugitive" part of the system, and this is the first time we have seen this used in the materials discussed. The solvent enables the user to deposit a thin film of a solid or viscous material; once its task is completed, the solvent leaves the scene. The resultant thin film—conformal coating—is tough, transparent, and hopefully, pinhole-free.

Epoxies, urethanes, and silicones cross-link on cure, as the solvent evaporates, offering enhanced chemical, solvent, and heat resistance. The enhanced resistance makes them more difficult to remove for component replacement or repairs. Acrylics, on the other hand, do not cure, but only dry. The solvent evaporates to leave a dry film, which can (if necessary) be redissolved with solvent for repair work. This improved "repairability," alas, is accompanied by poorer chemical resistance and much poorer solvent resistance.

Obviously, the solvent-containing materials are restricted to thin-film use, 5 mils or less per application, because the solvent must escape. Trapped solvent represents a "failed" application, so multiple coats are necessary.

Conformal coatings are applied by spray, dip, flow, or brush, the method selected being dictated by the number of boards to be processed, material costs, the uniformity of coating desired, and the availability of ventilation equipment. Dipping gives the most uniform coating.

Conformal coatings are not intended as the primary insulation on a PC board, for instance, but as protection against surface contamination, as a hold-down for components, and as a final line of defense when other encapsulants are used over them in potted units.

The materials most often encountered are acrylics, urethanes, epoxies, and silicone *resins*, not RTVs as previously encountered. A fifth and important conformal coating is used which does not fit the described methods of handling: Parylene,[*] a solid starting material, which is vaporized, then deposited on the substrate as a solid, and then cured. It will be discussed separately.

A. Acrylics

1. General Formula and Description

$$\left[-C - \overset{\overset{\displaystyle CH_3}{|}}{\underset{\underset{\displaystyle O}{\underset{\displaystyle |}{\underset{\displaystyle R}{|}}}{\underset{\displaystyle C=O}{|}}}{C} - \right]_n C + solvent \xrightarrow{\text{solvent } \uparrow} solid\ acrylic\ film$$

---

[*]Trademark of Union Carbide

Acrylic coatings are solvent solutions of an acrylic thermoplastic resin. No cross-linking (cure) occurs when they are deposited. The solvent evaporates, leaving a tough, hard, transparent film rather like clear fingernail polish. The dried film is virtually water-white, unless tinted, and has outstanding resistance to degredation by ultraviolet light. Acrylics, which operate to class B, have good general chemical resistance and poor solvent resistance, particularly to chlorinated solvents. Being a solvent solution of an unreactive plastic, they have an unlimited shelf life and, of course, no pot-life limitations beyond replacement of solvent lost through evaporation.

2. Advantages and Disadvantages

    Advantages

    High humidity resistance
    No shelf- or pot-life problems
    Good intercoat adhesion
    Very fast dry
    Very clear
    Can be soldered through
    Can be removed by solvent

    Disadvantages

    Very solvent sensitive
    Low solids—more solvent than plastic, necessitating multiple
        applications
    Expensive per mil of applied film; the evaporated solvent is
        lost money
    Strong solvent odor
    Solvents very flammable, with some now available in "no-
        flash" solvents
    Solvents may affect base board, components, or marking inks

3. Use Characteristics

Acrylics find use where fast drying with easy removal is important. Where high-humidity conditions are encountered, both short and long term, acrylics are a fine choice. The preferred method of application is dip coating because it gives the best coverage and most uniform coating thickness. Spraying is very wasteful of material. The board is dipped into a bath of coating at a speed consistent with thorough penetration and covering of all components as well as the board. Immersion and withdrawal speeds are critical and must be suited to the board being coated. Once the solvent has evaporated, a second coat, and pehaps even a third may be applied to assure a pinhole-free final conformal coating. Where repairs or component replacement are necessary, acrylics can be "soldered through" or removed with a solvent-soaked cloth.

B. Polyurethanes

1. General Formula and Description

$$O=C=N-R-N=C=O + \text{solvent} + \text{moisture} + \text{catalyst} \xrightarrow{CO_2 \uparrow \text{solvent} \uparrow} \text{cross-linked urethane}$$
(in air)

One-component moisture cure

$$O=C=N-R-N=C=O + \text{cross-linker} + \text{solvent} + \text{catalyst} \xrightarrow{\text{solvent} \uparrow} \text{cross-linked urethane}$$

Two-component

Urethane coatings are quite different from their relatives, the urethane potting materials, in that they are dissolved in a solvent and often are one-component materials (with some exceptions). As the solvent evaporates from the one-component system, the isocyanate groups react with moisture in the air. The moisture in turn causes the isocyanate to cross-link into a thermoset, hard, solvent- and chemical-resistant film, tightly adhering to the board components. The two-component urethanes, which require addition of a curing agent, behave much like an epoxy in that once mixed, they have a pot life and must be applied before thickening has progressed too far. *But* like their potting counterparts, urethane conformal coatings must be tightly sealed in their containers and protected from all moisture.

2. Advantages and Disadvantages

Advantages

Fast dry to touch, particularly the one-component moisture
    cure
Good heat resistance
Good chemical resistance
Good solvent resistance
Good electrical properties
No pot-life worries (one-component type only)

Disadvantages

Intercoat adhesion fair
Low solids
Expensive per mil of applied dry film
Moisture sensitive: potential shelf-life problems
Multiple coats to get desired film build
Strong solvent odors
Can "revert" in certain use conditions
Difficult to repair

3. Use Characteristics

Urethane conformal coatings are usually first choice where
humidity resistance over an extended period of time is of importance. Their chemical and solvent resistance are a drawback when
repairs must be made. They can be soldered through but leave a
brown discoloration behind. Extreme care must be exercised in
surface preparation prior to application, because even very small
amounts of moisture can cause problems of blistering. Urethanes
are applied much like the acrylics and with the same equipment.

C. Epoxies

1. General Formula and Description

$$\text{cross--linker} + \text{catalyst} \xrightarrow{\cong 320^\circ F} \text{thermoset epoxy}$$

The chemistry of epoxy conformal coatings is exactly the same as that for the room-temperature-cure potting materials. In most cases, the epoxy (base) portion is a solid rather than a liquid, but the cure chemistry is the same. Being a solid, these types must be dissolved in a solvent to permit applications.

All epoxy conformal coatings are two-component, and once mixed, the pot life must be adhered to. Epoxies are applied by the same techniques and equipment as urethanes and acrylics. Since epoxy coatings *can be* 100% solids, fewer applications would be required to give the desired film build—saving time and money since lost solvent is lost money. However, their dry-to-touch time is slower than those of acrylics and one-component urethanes.

2. Advantages and Disadvantages

Advantages

High solids (to 100%)
Good heat resistance
High abrasion resistance
Excellent solvent and chemical resistance
Economical per mil of applied dry film
High film builds
Infinite shelf life
Excellent adhesion to previous coats and to substrate

Disadvantages

More difficult to repair
Pot life
Slow cure
Lower long-term humidity resistance

3. Use Characteristics

The two components of the epoxy are mixed and handled the
same as a potting material, but since slow-curing agents are used, a
longer pot life results. In the case of coatings based on solid
epoxies, a pot life of 24 hr or more is possible. Application is the
same as for acrylics and urethanes and ultimate cure is measured in
days. Epoxy conformal coatings are preferred where tough, robust
coatings with good solvent and chemical resistance are desired, and
where high film build with minimum applications is an advantage.

D. Silicones

1. General Formula and Description

$$\left[\begin{array}{ccc} CH_3 & CH_3 & CH_3 \\ | & | & | \\ -Si-O-Si-O-Si- \\ | & | & | \\ CH_3 & O & CH_3 \\ CH_3 & | & CH_3 \\ | & | & | \\ -Si-O-Si-O-Si- \\ | & | & | \\ CH_3 & CH_3 & CH_3 \end{array}\right]_n + \text{catalyst} + \text{cross-linker} + \text{solvent} \xrightarrow{\text{solvent}\uparrow} \text{cured silicone film}$$

The silicones used in conformal coatings are different from the
RTV silicone rubbers used in potting applications, being harder
and more resinous in character, and are referred to as silicone
resins. They provide good humidity resistance and resistance to
corrosive chemicals. Their physical strengths are inferior to those
of epoxies, urethanes, and acrylics, and should not be relied on for

hold-down of components. Silicone resins provide some cushioning effect for delicate components, but their major asset is high temperature resistance—to 200°C. They are difficult to remove for repairs and do not vaporize, making solder or burn-through impossible. They must be cut and peeled away to effect repairs.

2. Advantages and Disadvantages

Advantages

Service to 200°C
Excellent resistance to long-term thermal degradation
Excellent humidity resistance
Excellent chemical resistance

Disadvantages

Very expensive
Very difficult to remove
Slow cure to ultimate properties
Low adhesion

3. Use Characteristics

Silicone conformal coatings are used where thermal performance approaching 200°C is required and where superior humidity and corrosion resistance is necessary. Their repairability (or lack of it) restricts their use to areas where the foregoing attributes are critical. They can be applied like the other conformal coatings, but usually are dip-coated.

E. Summary of Class III Materials

Property	Acrylics	Epoxies	Urethanes	Silicones
Pot life	Excellent	Fair	Good	Fair
Ease of application	Excellent	Good	Good to excellent	Good
Repair	Excellent	Fair	Fair	Poor

Property	Acrylics	Epoxies	Urethanes	Silicones
Abrasion resistance	Fair	Excellent	Good	Good
Temperature resistance	Poor	Good	Good	Excellent
Moisture resistance				
Short term	Excellent	Good	Excellent	Excellent
Long term	Good	Fair	Excellent	Good to excellent
Chemical resistance	Poor	Excellent	Good	Excellent
Cost, applied and dry	Fair	Good	Fair	Poor

III. PARYLENE

The Parylene process of conformal coating is a special case. Although the process starts with a solid, as in most class III materials, the process from this point on is much different

A. General Formula and Description

Di-para-xylylene (Dimer)   para-xylylene (Monomer)   Poly(para-xylylene) (Polymer)

Diagram of the Parylene process

Parylene is not practical to melt or dissolve, so it must be handled and applied by special equipment. As a starting point, it is a solid dimer which when applied becomes a thin, hard, tough, resistant, and very uniform coating over all components and the board, even sharp points and edges. The uniformity of the Parylene coating is remarkable, but unfortunately, it is very difficult to remove. It is very temperature resistant as well as resistant to chemicals and solvents. Uniform, pinhole-free coatings to 0.005 mil in thickness (approximately $1/10$ μm) can be applied.

B. Advantages and Disadvantages

Advantages

Very thin pinhold-free coatings can be applied
Absolutely uniform film applied over all components, in-
    cluding sharp edges and points
Only one coat necessary
Excellent solvent and chemical resistance
One-component system
Excellent moisture resistance
Excellent electrical properties

Disadvantages

Very expensive materials and equipment
Very difficult to remove for rework

C. Use Characteristics

Parylene can be applied only by special equipment. The solid
starting material, polyparaxylylene, is heated under vacuum and
disassociates to the monomer units. The temperature is then raised
more and the vacuum intensified to activate the monomer units.
This vapor is then deposited at *room temperature*, and under even
more intensified vacuum on the boards, components, and so on, to
be coated. The resultant coating is thin, uniform, tough, and resistant (and controllable). It is used where a final barrier to
moisture and chemical attack is critical and worth the high cost of
materials and equipment.

APPENDIX: SOURCES FOR MATERIALS

Solid Epoxy Resins

Ciba-Geigy, Ardsley, New York
Dow Chemical, Midland, Michigan

K & S Industries, Weston, Connecticut
Reichhold Chemical Co., North White Plains, New York
Shell Chemical, Houston, Texas

DAP

F.M.C. Co., Philadelphia, Pennsylvania
Hardwicke Chemical Co., Elgin, South Carolina

Phenolic Resins

Celanese, Louisville, Kentucky
Durez Division, Occidental Chemical, North Tonawanda,
    New York
Georgia Pacific, Atlanta, Georgia
Reichhold Chemical Co., North White Plains, New York
Rohm & Haas, Philadelphia, Pennsylvania

Polyester Resins

Allied Chemical, Morristown, New Jersey
American Cyanamid, Wayne, New Jersey
A.Z.S. Co., Eaton Park, Florida
Cargill Corp., Minneapolis, Minnesota
Reichhold Chemical Co., North White Plains, New York
U.S.S. Chemicals, Pittsburgh, Pennsylvania

Solid Acrylic Resins

Du Pont de Nemours, Wilmington, Delaware
Rohm & Haas, Philadelphia, Pennsylvania

Parylene

Union Carbide, Danbury, Connecticut

4
Methods of Encapsulation

I. INTRODUCTION

There is no such thing as the "correct" way to encapsulate a component or device, only the way that works best for the user. Some methods may be faster, cleaner, and less labor intensive than others, but any method that works satisfactorily for the material user is correct.

There are many methods of applying protective insulating materials currently employed in the electronics industry. The most popular is the simplest. Mix two components in a container and pour into a device. The most sophisticated is probably that used to apply Parylene, which involves vaporizing a solid, pyrolysis of the vapor to a reactive monomer, then vacuum deposition and polymerization on the substrate under an intense vacuum. Both of these methods are "correct" to the task at hand—the protection of an electronic component or device with the selected plastic material.

II. CLASS I MATERIALS

A. Preliminaries to Encapsulation

If the selected encapsulant is to be poured or injected into the device, it will be contained in one of two ways: by pouring into a mold cavity containing the device, with the mold subsequently removed and reused, or it will be poured into a case or can containing the device, with the case or can becoming a permanent part of the device, acting as its outer shell and wearing surface. Although encapsulating in a cavity or shell is the predominate method for class I materials, it is by no means exclusive. A significant amount of dip coating and impregnation is done.

   If the encapsulation is being done in a shell, the next step is pouring. If, however, the encapsulation is being done in a mold, a preliminary step is required for most materials—application of a release agent to the inside of the mold to ensure easy release of the "potted" device. The release agent selected will be determined by two considerations: the materials of construction of the mold and the type of encapsulant selected.

1. Mold Releases

In physical form, mold releases can vary from liquids to pasty solids to finely divided powders, usually in the form of microplatelets. The liquid types are applied by spray, brush, dip, or even pour-in, pour-out. Whatever the method, complete coverage of the mold is essential, particularly when using epoxy encapsulants, with their excellent adhesion to most materials used in electronics. The geometry of the mold often dictates the method of application.

   Mold releases can build up on mold surfaces, progressively degrading, which leads to poor surfaces on castings and poor release. It is therefore advisable to strip molds down to the mold surface periodically. This is done with solvents, chemical strippers, or even buffing.

Silicones. Silicones, used widely in both fluid and grease forms, give excellent release of most encapsulating materials. However, do not use them to release RTV silicones, as they are compatible with RTVs and can cause surface softening of an already soft material, and poor release. Above all, do not use silicones to release cast RTV from RTV molds. Silicone fluids are most often sprayed on to keep the amount applied to a minimum. Silicone greases are used to seal around the edges of two-piece metal molds, particularly old molds which have worn or warped, and do not "match up" well, permitting leaks. The grease does not soften and flow if the mold is heated, yet is firm enough to restrain the contents of smaller molds, and can be removed easily by wiping.

Two-component RTV silicone rubbers are also used to seal metal molds, particularly larger ones, with very prominent parting lines. They bridge the crack very well, stick well enough to present encapsulant leakage, do not soften under heat, release everything (except other RTV silicones), yet strip off easily after encapsulant cure. Spray silicones are recommended for release of epoxy from metal molds where high-temperature-cure systems are used.

Teflon* Spray. The Teflon (usually Vydax*) exists as tiny platelets suspended in a carrier solvent, which when applied and dried, overlap each other and prevent the casting material from touching the mold. Since they are discrete particles, care must be exercised to ensure a coherent coating. A "spray-buff-spray" technique is recommended where possible, with frequent shaking of the spray can. Teflon sprays are effective on smooth metal molds.

Vaseline.† Vaseline is an effective release agent, particularly for RTV silicones (silicone from silicone releases). When applied by wiping or brush, it is preheated to approximately 125°F, at which temperature it is very soft. It can also be dispersed in a solvent such as trichloroethane for spray or brush application. Not

---

*Trademarks of Du Pont de Nemours Co.
†Tradename of Chesebrough

recommended for high-temperature cures. Vaseline tends to leave a "textured" surface when brushed, which will be mirrored in the casting surface.

Paste Wax. A hard paste wax such as bowling alley wax or Butcher's wax is applied by wiping and should be lightly buffed after drying. Best used on plane surface molds, as it is difficult to apply it by wiping into intricate shapes or designs. Not recommended for high-temperature cures. Very effective for epoxies. Cure inhibition should be checked before using with silicones.

Pam. * Yes, from the supermarket. Releases most encapsulating materials. Useful on other than plane surface molds, as the spray reaches into areas where wiping is difficult. Also not recommended for high-temperature cures (greater than 150°F).

Silicone Pan Glaze. This is the baked-on silicone release coating used by bakeries. It is transparent and glossy and imparts a very smooth surface to the casting.

2. Molds Versus Shells

The economics of the manufactured device will dictate whether a mold or a can or shell is selected to encapsulate the device. A mold is reusable. Its cost per device encapsulated is variable, decreasing as the number of devices encapsulated increases. A can or shell, on the other hand, becomes part of the device and therefore becomes a component of the fixed cost.

General Considerations. Potting into a shell is very time efficient, with no mold "prep" time, no demolding, no casting cleanup time, and no mold reassembly—all with attendant labor. *However*, for some device configurations, an off-the-shelf shell may not exist, leaving a casting mold as the only alternative. The cost of a machined metal mold to injection or transfer mold the shell may exceed the cost of a few two-piece casting molds for the device.

If a plastic shell is selected, another point to consider is the adhesion of the encapsulant to the shell, particularly if thermal cycling is a part of the component specification. If free choice of

---

*Trademark of Boyle-Midway, Inc.

materials is permitted, stick with epoxy, phenolic, or DAP, as adhesion is usually excellent to all three. Some of the newer engineering thermoplastics are very difficult to adhere to. A case in point is polybutylene terephthallate; it is extremely difficult to get any potting material to stick well to it.

For very short or prototype runs, a temporary mold such as an RTV silicone or polyurethane rubber will probably suffice. Such a mold is very fast and easy to use, but has a short life, usually less than 100 pulls. An alternative is metal-filled epoxies, which are more difficult to prepare than silicones and urethanes, but have a longer life, greater than 100 pulls. *But remember*! Epoxy molds are *not* flexible and require a release agent, as do the urethane molds.

Permanent Mold Materials. Aluminum, steel, and Teflon are used as permanent mold materials. Aluminum and steel molds can be Teflon coated, which shortens and simplifies mold prep time. Teflon is easily scratched and damaged, however, so care must be taken, particularly when demolding. Only plastic tools should be used. Aluminum is easily more cheaply machined than steel, but can warp under some heat/cold conditions. Its light weight makes it more attractive as mold size increases. Steel and aluminum are considered to have infinite mold life if reasonably well cared for. Permanent molds should be periodically stripped clean of accumulated release agents to assure good cast part appearance and good release. Release of the casting from a metal mold can be initiated with a spatula inserted between the casting and the mold. If the mold has been permanently coated, this should be a plastic spatula to avoid damage.

Temporary Mold Materials. RTV silicones, polyurethanes, epoxies, and thermoplastics—polyethylene and polypropylene—are used. All are short run, probably 100 pulls or less, per cavity. All but silicones *require* a mold release agent, and even silicones can benefit from a release, giving improved life.

RTV silicone. Of this group, silicones give the shortest mold life, partly because of their lower physical strength, and partly because of their sensitivity to the encapsulants, particularly amine-cured materials. They are the easiest to work with, however, requiring no "prep" of the plug around which they will be cast, because of their inherent release properties. Their "easy" flexing makes demolding much easier.

A simple procedure can help extend silicone mold life. As the castings are released, note the point at which slight sticking begins. Wash the cavity out with a paint thinner or, better, trichloroethane, using a paintbrush. *Do not* soak the mold. Exposure should be limited to 5 min. Then bake out the mold at 150 to 200°F for 3 to 4 hr. Before reuse, lightly spray with a silicone release agent, then put back into service.

Polyurethane.   Urethanes, despite being flexible and not noted for their high-temperature properties, accept temperatures up to 200°F quite well and are being used for moderate-temperature-cure epoxies, with mold lives approaching 100 per cavity. Again, care must be exercised when releasing castings from hot molds to avoid cutting or splitting the material.

Polyolefin.   Polyethylene and propylene, being natural release agents, are used in sheet form, thermoformed into multicavity "gang" molds. After a few pulls, the surface tends to oxidize, and a release agent may be necessary at this point. If one cavity begins to give problems, just bypass it; continue to use the others.

Epoxy.   Metal-filled epoxies are best utilized for room-temperature curing systems, again with a release agent. Some special epoxies can tolerate elevated-temperature cures.

With all temporary mold materials, caution must be exercised when releasing the casting. With the flexible materials, break the casting away from one surface of the mold with the hands, preferably, then carefully insert an air line into a corner and flex the mold gently. With rigid molds, carefully insert a thin plastic spatula.

Expected mold life.   Following are the approximate number of parts per cavity.

Mold material	Casting material			
	Epoxies	Urethanes	Silicones	Polyesters
RTV silicone	10–25	50–100+	Not recommended	50–100+
Flexible urethane with release agent	50–100	100+	100+	50–100

	Casting material			
Mold material	Epoxies	Urethanes	Silicones	Polyesters
Metal-filled epoxy with				
release agent	50-100	100+	100+	50-75
Thermoplastics	25-50	50-100+	100+	25-50

One-Shot Mold Materials. A low-melting tin alloy is dip-coated over a "plug" of the correct size and shape, then cooled and removed. (Even high-melting jeweler's wax has been used.) This is used as the potting shell, and after the encapsulant has cured, is melted off. A question here is: Can the device withstand the melt temperature of the tin or wax? The one advantage is the reuse of the tin or wax, reducing the mold material cost to nearly zero. There is substantial labor involved, however.

One-Piece Versus Two-Piece Molds. Two-piece molds allow easier release of the castings since they disassemble. Application of the release agent is also facilitated. However, since there is a seam, leaks can occur, and a parting line on the casting is inevitable, and, if too heavy, some machining or cleanup will have to be done. Some material is lost in two-piece molds as "flash"—that material which flows into the mold seam and is cured there. It looks like a "ruffle" on the casting and must be removed.

Where an old or warped two-piece mold begins to leak seriously, silicone grease can be applied to seal it. RTV silicone rubbers are also used to seal. The adhesion is sufficient to prevent leaking, but they strip away very easily for part removal.

One-piece molds do not leak and give no parting line or flash, but make release and removal of the casting more difficult. If possible, the cavity should be designed with some "draft" to help the casting slip out. Even then, mechanical assistance is sometimes necessary, such as pushing a knife blade down one side to start the breakaway. Injecting compressed air into one corner is often helpful.

Whichever mold design is used, it is important that a release agent be selected which is appropriate to both the encapsulant being used and the mold material. Teflon and Teflon-coated metals, as well as RTV silicones, normally do not require release

(a)

Fig. 1 (a) A one-piece "gang" mold, here made from RTV silicone rubber. Photograph courtesy of Harvey Associates, Mt. Vernon, New York. (b) A two-piece metal mold in an open condition showing the molded coil. Photograph courtesy of Five Star Coil, Plantsville, Connecticut. (c) A two-piece metal mold in a closed position, with only the leads exposed. Photograph courtesy of Five Star Coil, Plantsville, Connecticut.

(b)

(c)

agents. Aluminum, steel, polyurethanes, epoxies, and thermo-plastics eventually do, and in the case of RTV silicones, a mold release helps prolong mold life (see Fig. 1).

B. Methods of Application

The best advice at this point—when a method must be selected—is: Keep it as simple as possible. While a meter-mix-dispense machine may sound appealing, work up to it in stages, learning as you go on. First, hand-mix and pour, then mix, degas, and inject with a cartridge or small pressure tank. Only if the throughput is still to low, or labor content is excessive, should a more sophisticated method be considered. With sophistication comes complication, as well as greater efficiency.

1. Hand-Mixing and Pouring

The simplest method of applying liquid encapsulants is the one most widely used. The two components are measured into a container, hand-mixed, then poured into the device or into the mold containing the device (see Fig. 2). If the system cures at room temperature, the job is finished. Just wait until it cures, usually overnight. If, however, bubbles rising to the surface cause a cosmetically unacceptable appearance, they can be broken by spraying the surface with an aerosol release agent such as silicone spray or, even cheaper, with a solvent. This lowers the surface tension and the viscosity of the surface of the bubbles, causing them to break and aiding in surface reflow. Safe work habits suggest using a chlorinated or fluorinated solvent, with their attendant high (or no) flashpoints. An inexpensive sprayer can be a pump hairspray bottle, suitably cleaned and filled with the appropriate solvent. Spray the surface very lightly. A second and even a third application may be necessary, depending on the amount of air in the encapsulant as received, and the amount mixed in with the curing agent. A slower-evaporating solvent is desirable, as it is retained on the surface and continues to break bubbles as they progressively rise to the surface.

Caution! This should be done only in a well-ventilated area!

Air bubbles trapped in the mass of the encapsulant are usually functionally undesirable, such as in high-voltage uses. Here,

Fig. 2 Hand pouring, the simplest application method. Photograph courtesy of Gamma High Voltage Research, Mt. Vernon, New York.

then, the first stage of sophistication is introduced. The mixed encapsulant must be degassed prior to being applied to the device.

Vacuum Degassing. The hand-mixed encapsulant is placed in a vacuum chamber and a vacuum drawn, enlarging and pulling the bubbles out. A vacuum desicator, or better, a plastic vacuum dome is usually sufficient (see Fig. 3). These domes are available in sizes up to several cubic feet in volume. One of these, an aluminum base plate, and a vacuum pump make an excellent and inexpensive evacuation chamber.

The two components of the encapsulant are mixed thoroughly, preferably in a wide container. The larger the surface area the better, to permit easier air escape. Also, the container should not be filled beyond the one-third mark. This shortens the distance the air bubble has to travel to escape and allows sufficient space for the "head" to rise and break without spilling over.

Fig. 3 A vacuum dome. It is shown on the base plate with a pressure gauge and valve. Mixed encapsulant is placed inside and evacuated. Photograph courtesy of International High Voltage Electronics, Danbury, Connecticut.

Ideally, the head will rise as the vacuum is drawn, then break near the top and collapse. Some bubbling will continue, but air removal is essentially complete a minute or two after the head collapses. Further evacuation accomplishes little, and, in some cases, can be injurious. See the caution remarks in the following section on vacuum potting.

This is usually the scenario with RTV silicones, but with epoxies and filled polyurethanes, it sometimes doesn't "follow the script". With these materials, the vacuum must often be broken to prevent the head of bubbles from overflowing. This partially collapses it, and helps break some bubbles. Repeating this procedure one or more times is usually effective in obtaining a bubble-free liquid. If the mixing container can be fitted with a mechanical agitator and operated under vacuum, the degassing stage becomes much easier and faster. The head rises very little and the bubbles break quickly. Even vibration is a help. The object is to keep the material in motion, which breaks bubbles faster and prevents excessive head buildup. A vibrating table is effective here.

In persistent cases, a commercial defoamer can be tried. It is mixed into the material after the curing agent portion is incorporated, and helps air bubbles escape. Since most of these are silicone based, an adverse effect on adhesion *may* be the result. Each case should be tested after seeking counsel from the encapsulant supplier.

The air in a batch of catalyzed encapsulant has two sources: (1) entrained during incorporation and dispersing the curing agent, and (2) in the encapsulant as received.

Air mixed in while catalyzing usually consists of a few fairly large bubbles, which come out fairly readily of their own accord, given time. Under vacuum, these large bubbles rise quickly and break easily. Mechanically mixing in the curing agent tends to aggravate this a little, and this should be a consideration. A high-speed dispenser should never be used for curing agent incorporation, as it introduces more air, in a finely divided form, which is much more difficult to remove.

The air contained in the encapsulant as received is much more difficult to remove. It is air that has been flushed out of the interstices of the filler particle, added to that introduced during the violent shear forces applied to the filler during its dispersion step. Whether this air escapes naturally or is retained depends on the type of filler used in the encapsulant. If the air is retained, it becomes quite difficult to remove, tending to rise into a semipermanent head, under vacuum, very reluctant to break. Agitation while under vacuum is one effective way of breaking these bubbles. Defoamers help, but surface sprays are not very effective, as the bubbles tend not to rise very rapidly. Preevacuation of the part A portion (part B, too, if it also contains filler) is another way to handle it.

When the device has densely packed components, just preevacuating the material may not be enough. Air can be trapped around and under the components and not be flushed out by the "poured-in" encapsulant. It then may become necessary to partially fill the device, then place it in the vacuum chamber and draw out this trapped air. The device is then "topped up" with encapsulant. Often, this can be avoided by the correct juxtapositioning of the components, then tilting the device or the mold in such a way that the air can flow around the components rather than be blocked by them. The liquid is then poured slowly, to flush out this air, rather than trapping it behind components.

The technique of introducing the material at the bottom of the can or mold through a funnel or the nozzle of a caulking gun can be very helpful in producing air-free units, and is often used with RTV silicones, because of their ease of air removal.

Place the end of the funnel or tip of the caulking nozzle on the bottom of the unit or mold cavity. Slowly introduce the pre-degassed encapsulant, allowing it to spread across, then up through the components, displacing the air ahead of it. If the unit can be tilted with the introduction point at the low end, it will further reduce the chance for air pockets.

Vacuum Potting. The final stage of "extreme measures" in vacuum encapsulating is to fill the device while under vacuum. Here the device is placed under vacuum, then the encapsulating material is introduced while the device is still under vacuum (see Fig. 4). When the device is full and the vacuum released, the air outside acts as an additional "ram" to push the encapsulant in. The encapsulant should be predegassed for this procedure. There are units commercially available to do this, as well as many home-made designs, the purpose of the entire exercise being to get a void-free insulating medium into the device.

A word of caution on vacuum. Many encapsulants can be adversely affected if held under vacuum too long. Key ingredients can be distilled out by the vacuum, with resultant inferior cure rate and in some cases, inferior cured properties. Maintain the vacuum for the minimum time required. If peculiarities in cure rate or properties are noted, contact the encapsulant supplier for advice and counsel. The cure rate of condensation-cure silicones, for example, is quickly affected by extended periods of evacuation of more than 15 min. Once the head has risen and broken, 1 to 2 min more evacuation is sufficient. More vacuum can cause these cure problems, so keep it as short as possible.

2. Mechanizing Potting

Injection. The first step in mechanizing the "pour-it-out-of-a-can" method of encapsulation is to pour the mixed material into a syringe or caulking cartridge and inject it into the device (see Fig. 5). This is effective in getting material into small or densely packed devices with a minimum of trapped air. The end of the needle or nozzle can be placed at the low point of the device, and as the encapsulant is forced out at the bottom, the air is displaced upward, without having to "bubble up" through the incoming material, resulting in much less trapped air. The syringe or caulking cartridge can be connected to a compressed-air source to speed the delivery of encapsulant and relieve operator fatigue.

Metering Parts A and B. The second step in automation is to meter the two components mechanically. Such equipment is designed to eliminate measuring errors. Each component is delivered simultaneously in a measured amount into a mixing container. This equipment does not mix; it simply meters out each component in its correct ratio (see Fig. 6). The procedure at this point becomes the same as outlined at the beginning: mix, pour, and/or evacuate, and/or inject with a syringe.

Meter-Mix-Dispense. The third and final step in mechanization is the meter-mix-dispense machine. If encapsulation is being done on a continuous basis, meter-mix equipment can eliminate a lot of errors, mess, and labor. This equipment not only measures out each of the two components in its correct ratio, but mixes and dispenses the mixed material through a nozzle which is inserted into the device. It is even possible to have a predetermined shot size with this type of equipment, which can be an advantage to eliminate overflow and subsequent cleanup problems when filling small devices (see Fig. 7).

Most meter-mix machines come with a fixed ratio, which is determined by the encapsulant selected. Two accurately machined cylinders each delivers the precise amount of each component to the in-line mixer. A little more expensive, but a great deal more useful, are variable-ratio machines. These permit the user to use

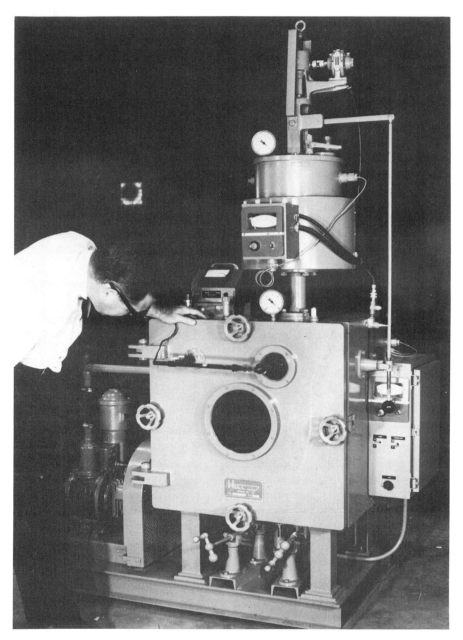

Fig. 4(a)

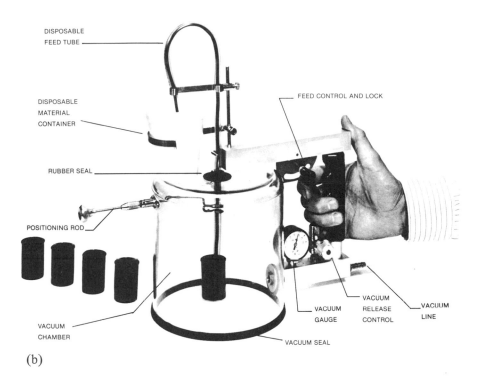

(b)

Fig. 4   (a) A production vacuum encapsulator. The upper chamber permits mixed encapsulant to be heated and agitated as well as evacuated. The lower (larger) chamber can also be heated. Photograph courtesy of Hull Corp., Hatboro, Pennsylvania. (b) A small, hand-held vacuum encapsulator. It can be placed over small units and quickly evacuated for small or prototype jobs. Photograph courtesy of Envax Products, Inc., Oxford, Connecticut.

(a)

Fig. 5   (a) Injection syringe. Mixed material is loaded into the syringe and injected by means of air pressure. Carefully regulated, very precise amounts can be dispensed. Photograph of Ashby-Cross (Danvers, Massachusetts) equipment courtesy of Methods Engineering, Vauxhall, New Jersey. (b) Injection on a larger scale. Mixed evacuated encapsulant is loaded into a 16-ounce plastic cartridge and dispensed by means of air pressure. Photograph courtesy of International High Volt Elect., Danbury, Connecticut.

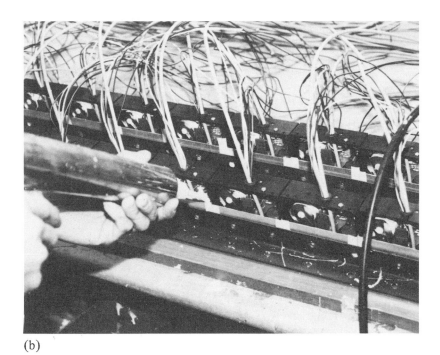

(b)

materials with different ratios merely by cleaning out the containers and lines and making the necessary adjustments. The higher initial cost is usually justified by the additional freedom.

When considering the incorporation of a meter-mix machine into manufacturing, the user should realize that it is not a panacea. It is only as good as the *least* capable person who will operate it. You are not eliminating your "potting" problems; rather, you are exchanging a myriad of unpredictable problems for a few more predictable, more managable problems. The predictable problems are of the type "press the wrong button and you get the wrong result."

Considerations on meter-mix-dispense equipment. What do you get when you spend $5000+ for a meter-mix-dispenser unit? Following are a few points to consider.

> Human measuring and mixing errors are eliminated, with resultant material and device savings. However, mechani-

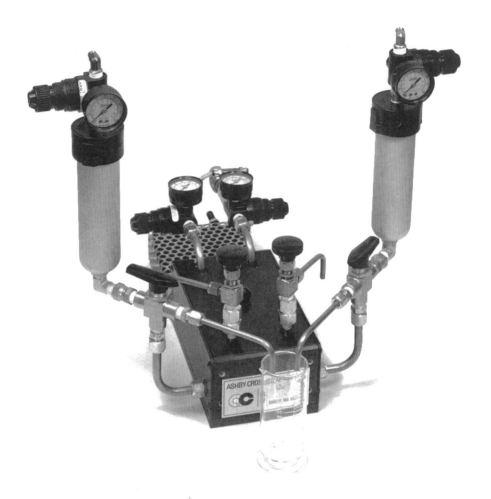

Fig. 6 Metering only. Each component is measured and simultaneously delivered into a mixing container. Photograph of Ashby-Cross Equipment courtesy of Methods Engineering, Vauxhall, New Jersey.

cal mixing and measuring errors can occur, but these are easier to trace; dispensers do not lie!

No air bubbles are added when the two components are mixed in the equipment, as happens in hand or power mixing. This eliminates the need to evacuate in many cases.

Since a two-component encapsulant has a pot life, that part
of the dispenser which contains mixed material (i.e., the
mixing head or tube) must be purged with fresh mixed
material or one of the unmixed components, to avoid
having the encapsulant gel in the equipment. The frequency of purging is, of course, dependent on the pot
life of the mixed encapsulant, at the dispensing temperature. This purged material is wasted and adds to the cost
of materials. This should be a consideration when considering whether to automate. A rule of thumb is to
purge when 60% of the pot life has been reached.

Each time the dispenser is started up after a shutdown, a
sample shot of mixed material should be taken and
cured, to confirm that everything is functioning properly. A gel time and hardness are usually sufficient.

More material can be dispensed into more devices by a dispenser than can be done by hand, and done more quickly.

Dispensers do not operate well in run-stop-run situations.
They are most effective when run continuously. Every
time a dispenser is stopped or on standby, material is
wasted.

On-the-Fly Degasser.   There is one last option that turns the
meter-mix dispenser into a complete device—a continuous "on-
the-fly" degasser can be added. This enables the user to deliver accurately measured, mixed, and evacuated encapsulant to the device. Under most circumstances, this eliminates the need for
further evacuation during encapsulation. However, this now complete device becomes completely expensive, with prices ranging to
$50,000. The vacuum is applied to the A and B parts separately
just before mixing, which does not add any additional air.

3. Dipping

When the job at hand is to apply a thin 1/16- to 1/4-in. protective
coating over a board, transformer, capacitor, and so on, there is
little equipment available; it becomes a hand operation. A meter-
mix machine is of little help, as the materials are thixotropic and
do not flow well enough to be handled in this equipment.

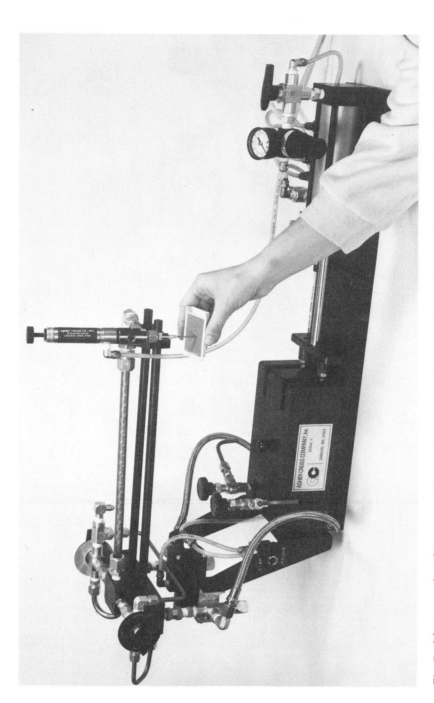

Fig. 7 (a) A meter/mix/dispenser. After each component is measured, it is fed into the upper mixing tube and dispensed as a mixed encapsulant. Photograph of Ashby-Cross equipment courtesy of Methods Engineering, Vauxhall, New Jersey.

Fig. 7 (b)Meter/mix/dispense equipment with both component reservoirs separately agitated, and enclosed in a constant temperature enclosure with an exterior solvent flush tank. Photograph courtesy of Methods Engineering, Vauxhall, New Jersey.

(a)

Fig. 8 Dip encapsulation. Transformers are dipped into the encapsulant, hung and drained, then cured. (a) Transformers before dipping, (b) dipped transformers on drain rack, (c) drain rack transferred to cure oven. Photos by Bill Flaim, courtesy Fernwood Transformer, Bellevedere, New Jersey.

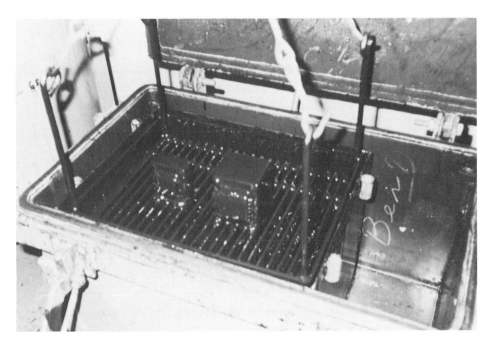

(b)

(c)

The device is dipped into a bath of thixotropic encapsulant, withdrawn, and hung to allow excess material to drip off (see Fig. 8). Often, the bath of dipping material is in a vacuum container. The devices are dipped; a lid is secured; vacuum is drawn then released; and the devices are removed. The vacuum-release sequence draws surface air off uneven surfaces, such as transformer laminations, and helps force material into windings, under components, and so on. Finally, the coated devices are hung for curing. The amount of material retained on the device is determined by several factors:

1. Viscosity and degree of thixotropy of the material.
2. Rate of withdrawal of the device out of the material.
3. The temperature of the device as it is dipped (the warmer it is, the less it picks up, as it lowers the viscosity of material touching it, causing more flow-off).
4. Vibration of the dipped device will cause additional run-off.
5. Use of a solvent in the dipping material can be another control, as it lowers viscosity and lessens pickup. It represents a risk, however, as it can be trapped in a thick coating and cause cure problems and in the case of heat-cured systems, popping and blow holes. Long-term weight loss and shrinkage are also risks, as retained solvent will slowly evaporate out of the coating.

4. "Buttering On" as a Variation on Dipping

When units become too large to be readily moved around, and a coating is required, a simple expedient is the use of a nonflowable thixotropic paste—exhibiting less flow than materials used in dipping—which is, literally, "buttered on" with a spatula (see Fig. 9). These encapsulants are usually room-temperature cured, as the large component cannot be moved into an oven because of its size. The technique lends itself to mixing up several small batches, so pot life is not usually a problem.

Fig. 9 Buttering-on. Here a stiff paste brush is used to apply the thick pasty encapsulant to a transformer. The leads are tied to a rod, up and out of the way. Photograph courtesy of Kip Transformer, Wolcott, Connecticut.

5. Impregnation

Certain devices, such as coils, transformers, capacitors, and so on, require that a liquid dielectric be impregnated into the windings. The smaller the gauge of wire or capacitor winding, the more difficult it becomes. A prerequisite for this procedure is that the impregnant be very low in viscosity, usually 1000 cP or less. This, then, pretty much eliminates the use of filled materials, as the increased viscosity and the filler particles inability to get past any but the coarsest windings make unfilled liquids the only choice [1]. Originally, solvent-thinned varnishes, often containing more solvent than resin, sufficed and still do for many uses. The solvent

(a)

Fig. 10 Impregnation. (a) Transformer prepared for impregnation. (b) The vacuum tank. (c) Draining after impregnation. Photographs courtesy of Del Electronics, Mt. Vernon, New York.

(b)

(c)

lowered the resin viscosity and improved inpregnation. But the lost solvent became void areas between windings, and as more complete fills became necessary, for thermal and voltage reasons, and windings became finer and denser, solvents—the great impregnators—began to disappear. Impregnants of 100% solids, with attendant higher viscosities, became necessary. Getting them in became an increasing problem. Time and elevated temperature helped but were not the solution. The solution became vacuum impregnation, also used now with solvent systems to speed up their impregnation (see Fig. 10).

6. Vacuum

Vacuum impregnation or potting means just that—applying the liquid encapsulant while the device is under vacuum, or applying the vacuum immediately after the impregnant has been applied or the device placed in the impregnant. The use of the vacuum is an "add-on" technique that can be attached to any of the foregoing methods.

The simplest method is to fill a potted unit partially, place it in a vacuum chamber, draw out the air, then top-up with more material. A better and more successful technique is to arrange a port through the vacuum chamber, through which a tube passes into the unit. Now, once the unit has been evacuated, the material can be introduced into the device while it is still under vacuum, without the exiting air having to bubble up through the material, with subsequent frothing. Faster, better impregnations result. The final shove comes when the chamber is vented to the atmosphere. The incoming air completes the filling process by imposing atmospheric pressure on the now hopefully full device.

In the case of encapsulated transformers, vacuum is recommended even if filling of the windings or corona prevention are not considerations. Most dip encapsulants are heat cures, as potlife considerations dictate this. The air trapped between the laminations will expand when heated during the cure cycle and often causes "pops" or blow holes in the skin of the encapsulant. A brief exposure to vacuum after the transformer has been immersed quickly pulls this air out, replacing it with encapsulant when the vacuum is broken, and illuminates these flaws. The transformers are then hung to drain, then oven cured.

The same technique applies to varnish impregnating coils, capacitors, and transformers. Small units can be placed in a bath of impregnant and evacuated, then hung to drain and cured. With very small units, a basket can be used to hold large quantities just tumbled in, then impregnated as above. They would then be individually hung, or placed in racks, for subsequent oven cure.

7. Pressure

The best way to assure void-free castings is by the use of pressure rather than vacuum. Unfortunately, it is impractical to rely on pressure alone, as it must be maintained throughout the whole cure cycle, whereas vacuum does its job during application, then is done with. If the pressure is released too soon, the bubbles will reappear.

Briefly, the unit is filled with encapsulant at atmospheric pressure. Provision is made for a resevoir of fresh encapsulant to be available once the pressure is applied. Fifty to sixty pounds of air pressure is applied in a pressure tank, and all bubbles begin to collapse, eventually disappearing. This is the reason a source for fresh encapsulant is necessary to replace the volume of the lost bubbles. Absolutely void-free castings can be made repeatedly this way, whereas even very intense vacuum usually leaves some small residue air or a void somewhere. So although "vacuum" is second best, it is good and much more practical, as it is applied for a short period of time, then finished.

8. Other Methods

There are several other methods of application of class I encapsulants. They are in infrequent use, but as references are occasionally made to them in the trade press, it is useful to be aware of them.

Pressure Gelation. A barrier to the rapid production of medium to large castings has been the need to heat the casting material to very high temperature, 150°C and above, with resultant runaway exotherms which cause distortion of and cracks in

the castings. Differential stresses are produced by the expanding
hot spot in the fat part of the mold.

It has been established that most of the shrinking that occurs when liquid thermoset systems cure takes place in the liquid
state prior to gelation [2]. Gelation occurs first in the widest and
thickest part of the mold, where the exotherm occurs first and is
highest. Once the gelled portion has grown to the point that the
mold is in effect divided, the other ungelled areas, such as edges
and corners, no longer have access to a fresh supply of liquid ungelled material to compensate for the shrinkage that occurs during
cure. It cannot get past this gelled "seal" in the fat part of the
mold. Thus pull-away (cavities) and stress are built up.

Pressure gelation uses higher-reactivity epoxy-anhydride systems and preheats the mold to a temperature of at least 50°C
higher than that of the resin portion, which is pressured in at a
temperature of 80 to 120°C. This, together with 25 to 35 lb of air
pressure applied to the resin as it is fed into the mold, make up the
pressure gelatin process (see Fig. 11).

Gelation now occurs from the heated walls of the mold in-
ward and from the farthest parts into the entry point. Remember,
the resin is relatively cooler than the mold, so the material nearest
the inlet will be the coolest. Now the majority of the shrinkage occurring while the resin is still liquid can be fully compensated for
by the entry of fresh material.

Relatively stress-free gel castings are the result, at cycle times
of 10 to 30 min, reduced from 3 to 8 hr for conventional liquid
casting techniques. Once demolded, these fast-gelled castings are
postcured outside the mold, making it available immediately for
another 30-min cycle.

Pressure gelation is used to produce medium to large castings
of high-voltage switchgear, bus bar units, power circuit breakers,
switches, and so on. Obviously, metal molds are necessary, as flexible molds would become distorted and deteriorate very rapidly.
This technique is used more widely in Europe than in North
America.

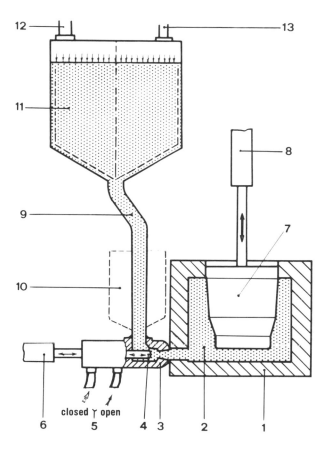

Fig. 11 Schematic diagram of pressure gelation. 1—mold, 2—casting, 3—casting valve, 4—plunger, 5—lines for compressed air that actuates the plunger, 6—pneumatic cylinder that moves casting valve to and from the mold, 7—core, 8—core pulling cylinder, 9—casting mix feed line, 10—heatable intermediate tank fitted when casting mix requires preheating, 11—pressure tank, 12—feed line from mixing vessel, 13—line connecting tank to compressed air supply or vacuum pump. From Ciba-Geigy publication.

Liquid Injection Molding.*   As the name implies, a liquid is injected into a mold, where it cures. The equipment is a modified injection molding machine such as would be used to mold thermoplastics or the molding powders of class II.

Two things mark liquid injection molding (LIM) as different from powder injection or transfer molding:

1. Powders are heated and melted in the screw section, whereas liquids do not require this, the screws serving only to inject and sometimes to mix the two components.
2. Cycle times run to 10 to 15 sec compared to 2 to 4 min for thermosetting powders.

To date, mainly special silicone polymers (not RTVs) have been used in this technique, and primarily to mold rubber components, with some encapsulation of components with simple configurations (see Fig. 12).

Centrifugal Casting.   Centrifugal casting involves injection of a liquid casting material into a mold contained in a rotating cylinder. Centrifugal force presses the casting material down into the mold and forces air out. Very fast curing liquid systems can be handled with this technique without most of the problems of complete mold filling with normal mold-filling techniques.

Very fast gelling urethane and epoxy systems are being used to mold very small parts and components such as connectors, where the cost and part volume does not justify transfer or injection molding.

A very large high voltage component currently being molded by a variation of this technique is described in Ref. 4. The parts in this case weigh 45 lb and are molded and removed on a 1-hr cycle. The actual gel time is 20 to 30 minutes.

The molds are heated or not as the situation demands. Where the molds are heated to temperatures above that of the casting material, you are, in effect, borrowing from the previously described pressure gelation technique (see Fig. 13).

―――――――――――
*For more information, see Ref. 3.

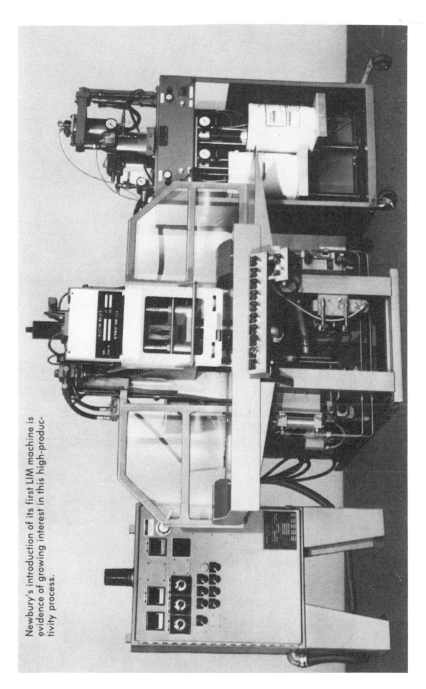

Newbury's introduction of its first LIM machine is evidence of growing interest in this high-productivity process.

Fig. 12 Liquid injection molding — L. I. M. Photograph from Plastics Technology, May, 1983.

Fig. 13 Centrifugal casting machine. Photograph from Plastics Technology, September, 1974.

III. CLASS II MATERIALS

The utilization of powdered encapsulants usually necessitates the expenditure of considerable capital, as molding equipment is expensive. Just the molds for transfer, injection, and compression molding cost are in excess of $10,000 and often $25,000 to $50,000, and one is required for each device to be encapsulated. Why use a molding powder when a device can be encapsulated with a liquid system for minimum capital expenditures? Compression, injection, and transfer molding lend themselves to high-speed production of components having a simple configuration. The resultant encapsulated component looks good, the surfaces mirroring the mold surface, and is produced in a matter of a minute or two. If large quantities of the same device must be produced, the high initial equipment and mold costs can be written off quickly, and the cost per part can become very low.

FAIRCHILD RESEARCH
TECHNICAL INFORMATION CENT

A. Compression Molding

1. Description

Compression molding was the original way to mold powdered plastic materials.* It consists of a high-pressure press and a precision-machined two-piece-metal mold (see Fig. 14). The top or "force plug" of the mold moves up and down. The plastic powder or powder preform is placed in the bottom, fixed cavity of the mold. The mold is heated and the top moves down, enclosing the change within the closed mold. Heat and pressure are increased, to 5000 to 10,000 lb, until the powder melts, flows into all parts of the mold, and cures (sets). This takes about 2 to 4 min. The two halves of the mold are then separated and the finished part or component is removed from the cavity.

Since inserts are difficult to mold around, compression molding is usually used to mold actual parts, such as coil bobbins, potting shells, and wiring devices. Connectors are molded by this technique. A full coil would be molded more easily by transfer or injection methods.

2. Advantages and Disadvantages

Advantages

Low material waste: no runners to be filled and wasted
Casting cures to a uniform condition; with little flow involved, there is little mastication of the charge with subsequent filler breakage and fiber orientation
No gate marks as with transfer molding
Simpler molds
Lower-cost molds
Less erosion of the molds by the charge

---

*Thermoset molding–SPI (see *Modern Plastics Encyclopedia*, 1980 (1981).

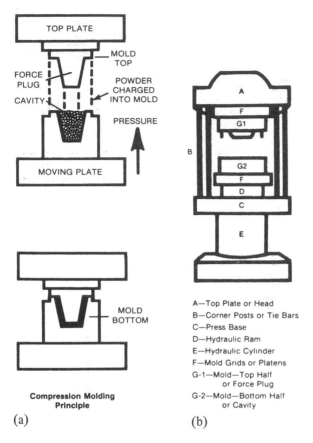

A—Top Plate or Head
B—Corner Posts or Tie Bars
C—Press Base
D—Hydraulic Ram
E—Hydraulic Cylinder
F—Mold Grids or Platens
G-1—Mold—Top Half
　　　 or Force Plug
G-2—Mold—Bottom Half
　　　 or Cavity

(a)   (b)

Fig. 14   (a) Simple schematic diagram of Compression Molding Process. (b) Components of a Compression Molding Press. Figs. a and b from "Thermoset Molding", Society of the Plastics Industry, Inc., New York, New York. (c) Commercial compression molding press. Photograph courtesy of Hull Corp., Hatboro, Pennsylvania. (d) A compression mold. Note thick steel parts to withstand the tons of clamp pressure. Photograph courtesy of Endicott Coil Co., Binghamton, New York.

(c)

(d)

Disadvantages

Difficult to fixture an insert
Casting has a heavy mold parting line
Complex shapes can result in incomplete mold filling
Multicavity molds cause slow mold cycles and inconsistent
      fills and densities unless each cavity is separately filled
      with powder or a preform
Powder preform should be preheated to shorten cycle times
High capital cost of molds and press

B. Transfer Molding

1. Description

There are several variations of the basic original transfer molding
method. Basically, the charge (powder or compacted powder preform) is placed in a preheat chamber in the center of the mold, the
cavities of which radiate out from the central preheat chamber and
connected to it by canals or runners. The mold halves are clamped
together and a plunger or ram transfers or pushes the now molten
material out of the pot, along the runners into the cavities. The
cure is complete in 1 to 2 min. The mold is opened, parts ejected,
runners and flash cleaned off, and the cycle is repeated.

Since inserts are easily molded around in transfer molding, it
is often the choice for molding coils, resistors, and connectors.
These components are placed in the cavities prior to mold
clamping.

2. Advantages and Disadvantages

Advantages

Many small components can be rapidly encapsulated because
      of low melt viscosity and "ram" pressure.
The encapsulated components have precise dimension.
Cures faster than compression molding since cure begins in
      the preheat chamber.
There is less breakage of pins and cores.

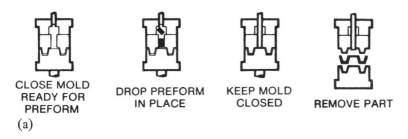

CLOSE MOLD
READY FOR
PREFORM
(a)

DROP PREFORM
IN PLACE

KEEP MOLD
CLOSED

REMOVE PART

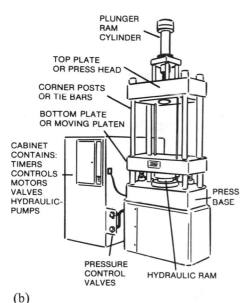

PLUNGER
RAM
CYLINDER

TOP PLATE
OR PRESS HEAD

CORNER POSTS
OR TIE BARS

BOTTOM PLATE
OR MOVING PLATEN

CABINET
CONTAINS:
TIMERS
CONTROLS
MOTORS
VALVES
HYDRAULIC-
PUMPS

PRESS
BASE

PRESSURE
CONTROL
VALVES

HYDRAULIC RAM

(b)

Fig. 15 (a) Simple schematic diagram of the transfer molding process. (b) Components of a transfer molding press. Figs. a and b from "Thermoset Molding", Society of the Plastics Industry, Inc., New York, New York. (c) Commercial transfer molding press. The vertical "rod" in the enclosed area is the ram, which moves the molten charge through runners into the cavities. Photograph courtesy of Hull Corp., Hatboro, Pennsylvania. (d) Photographs of a "four up" transfer mold open and installed in a press, courtesy of Endicott Coil Co., Binghamton, New York.

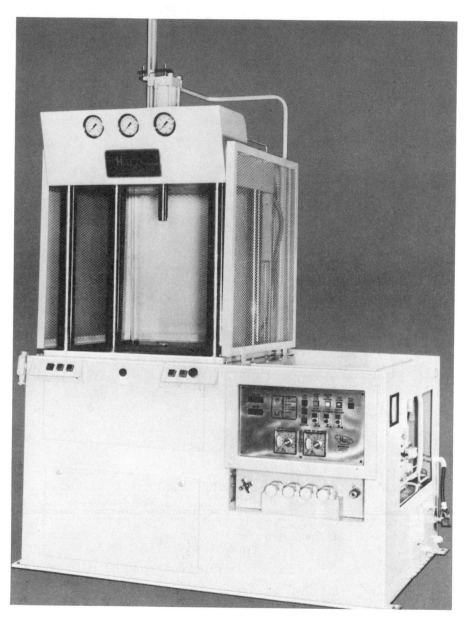

Fig. 15 (c)

(d)

(e)

Cycle times are faster than compression, as a single charge
services all of the cavity.

There is less wear on the mold parting line because there is
less flash.

Inserts are more easily molded in.

There is uniform density of molded pieces.

Disadvantages

Powders are more expensive than for compression.

Material in runners is wasted, contributing to higher material
cost per component.

There is high capital cost of mold and press.

Mold gates make mold more expensive than compression.

Fiberous fillers become "oriented" and even broken as
molten material is pushed down the runners, and particularly through the gates. This can show up as differential shrinkage and physical properties which vary between the flow and cross-flow directions.

Gate location is critical.

Large flat parts often show flow lines.

Part size is limited.

The powder or preform should be preheated to shorten the
cycle time.

C. Injection Moldings:* Reciprocating Screw

1. Description

In injection molding, the powder change is fed into a screw/ram
housed in a sleeve that is heated. As the screw rotates, material is
forced forward, picking up heat from the heated sleeve or barrel
and from friction as it passes over and around the rotating screw
vanes. As it heats up, it becomes more fluid. As more material is
deposited at the top of the screw, the screw, is forced backward

---

*For more information, see Ref. 5.

against a spring or hydraulic cylinder. At a precise point in its movement backward, the screw stops turning, then is forced forward by a hydraulic cylinder and the molten material is forced through the sprue and runner network into the mold cavities. Injection molding exhibits all the advantages of transfer versus compression molding, so components can be encapsulated as in transfer, but done much more quickly (see Fig. 16).

2. Advantages and Disadvantages

Advantages

Can do all the jobs of transfer molding.
Has faster cycle than transfer.
Less labor intensive than transfer, as the screw is fed from a hopper.
No preheater required.
No preform required.

Disadvantages

Most mineral-filled material unsuitable because of screw abrasion. The screw a very expensive part to replace.
Difficult to clean out if a plug up occurs.
Pressure control is crucial to prevent flashing.
Capital cost is high.

D. Fluid Bed Coating

1. Description

There is one final method of encapsulation, a fluid bed, which makes use of a powdered encapsulant. It is simply a small "dust storm" of the powder created in a cylinder. The finely and uniformly ground powder contains many of the same ingredients as those of an epoxy transfer compound. It is placed in a cylinder with an air diffusion plate at the bottom. Low-pressure air is passed through the plate and aerates the powder so that it rises,

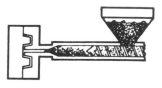

1. CLOSE MOLD.

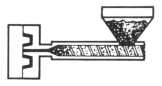

2. SCREW MOVES FORWARD TO INJECT HOT MATERIAL.

4. SCREW ROTATES, MOVES BACKWARD, HEATS MATERIAL.

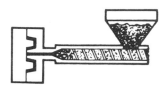

3. MATERIAL CURES IN CLOSED MOLD.

(a)

5. MOLD OPENS, EJECTOR PINS KNOCK OUT PARTS.

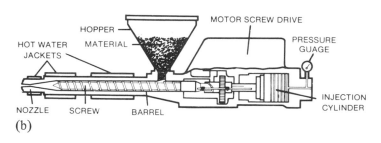

(b)

Fig. 16   (a) Simple schematic diagram of the injection molding process. (b) Components of an injection molding machine. Figs. a and b from "Thermoset Molding", Society of the Plastic Industry, Inc., New York, New York. (c) Commercial Injection Molding Machine. Photograph courtesy of Hull Corp., Hatboro, Pennsylvania.

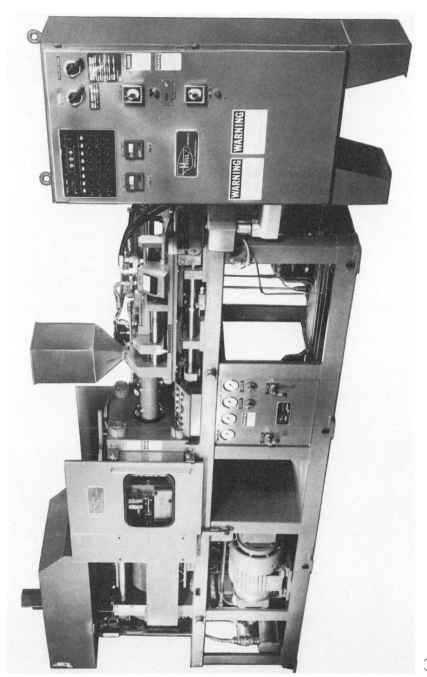

and the particles assume a random dust storm motion. The component—coil, capacitor, toroid, and so on—is preheated to a temperature above the sintering temperature of the epoxy powder. It is then dipped into the dust storm and slowly removed. The particles that touch the hot component soften, stick, and fuse into one another and the component. The coated component is then placed in a postcure oven to complete the sintering and cure of the coating.

Fluid bed is a fast, effective way to put a medium-thick (25 to 100 mils), tough, hard coating over coils, toroids, powdered cores, and the like. Thermoplastic powders can also be used. Any post-temperature treatment of the thermoplastics is very short, just completing the sintering, so no cure is necessary (see Fig. 17).

2. Advantages and Disadvantages

Advantages

One-component systems used.
Very high molecular weight polymers can be used without viscosity problems, thus contributing to toughness.
Very modest capital expense.
Easy to start up and shut down.

Disadvantages

Components must be able to withstand high preheat and even higher postcure temperatures.
Difficult to get a pinhole-free coatings.
Powders are expensive.

E. Electrostatic Spray

One drawback of fluid bed coating is poor edge and point coating. As the powder melts, surface tension tends to pull it away from edges and points. In electrostatic spraying, the powder is electrically charged. This charge actually intensifies at corners and edges so that the powder covers these extremely well. Also, less pinholing

Fig. 17 Fluid bed. Photograph courtesy of ARTTED Co., Springfield, Massachusetts.

is evident. An added advantage is even although it is a spray technique, the charge draws the powder toward the object, so little is lost to overspray.

IV. CLASS III MATERIALS

Very little by way of equipment is necessary to apply the conformal castings of class III. They can be applied by dip, brush, spray, or flow coating. The most important consideration is that the coating be applied uniformly and free of pinholes.

A. Dip Coating

1. Description

Dipping is the most efficient method and the most effective route to uniform coatings. The speed of immersion of the substrate is very important, as air must be flushed off the surface and from around and under components. Too-rapid immersion will overlap and trap air in pockets. If large numbers of boards or components need to be dipped, an automatic process should be considered which dips and removes the boards according to a predetermined schedule.

2. Advantages and Disadvantages

Advantages

Uniform coating thickness
Least number of bubbles

Disadvantage. Rapid evaporation of solvents from the coating bath causes variations and must be monitored carefully with makeup solvent added.

B. Spraying

Spraying is the fastest method but most wasteful of material. A spray booth is recommended, as most of the solvents are flam-

Fig. 18 Automatic conformal coating spray equipment with integral drying oven. The automatic feature helps improve coating uniformity and reduce overspray. Photograph courtesy Integrated Technologies, Acushnet, Massachusetts.

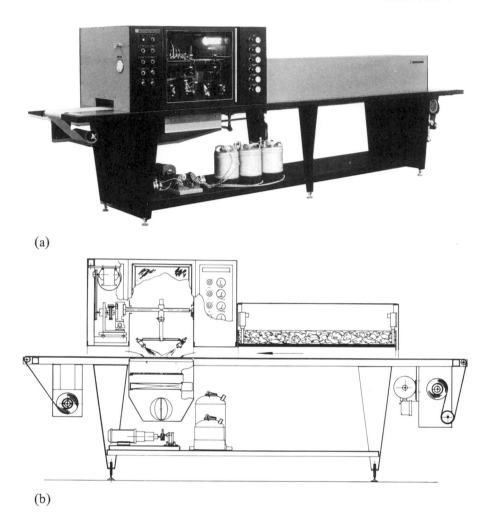

(a)

(b)

Fig. 19 (a) and (b). Automatic sprayer with angled nozzles for maximum coating and a disposable paper belt. Photographs courtesy of Integrated Technology, Inc.

mable and all are considered hazardous, so breathing should be minimized.

Spraying results in rather nonuniform coatings and often blows air in under components. Once an air bubble is "dried in," it is perpetuated through subsequent spray applications.

C. Brushing

Brushing makes uniform coatings very difficult but is often used in prototyping for reasons of expediency. It is quick and easy for an occasional board. High-quality brushes must be used, as stray bristles left behind will provide a convenient moisture path into the board.

D. Flow Coating

Flow coating is really a variation of dipping. It involves pouring coating down over the board, which is held at a 45° angle over a "catch" trough. It results in puddling and very nonuniform coatings, and is used mainly with very large substrates to replace brushing for prototype and short-run jobs.

V. PARYLENE

A. Description

Parylene can be applied only by expensive, very specialized equipment, so its use demands a high part throughput, unless its cost can be justified for research purposes or very high reliability (and expensive) military aerospace equipment.

Unlike the solids in class II, the solid Parylene starting material, a dimer, does not melt. It is a powder that is deposited from, and in, a vacuum [6]. It begins with the vaporization of the powdered solid at approximately 170°C in the evaporation zone. Pyrolysis of the dimer results in the breaking of the two methylene bonds (see Section IV of Chapter 3) at about 675°C, yielding

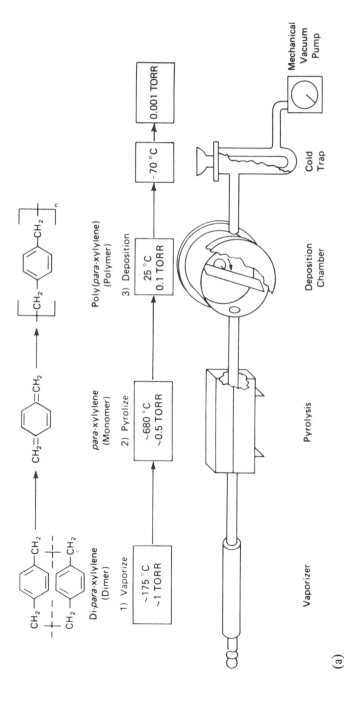

Fig. 20 (a) Schematic of a commercial parylene coating maching. Courtesy of Union Carbide. (b) Commercial parylene coating machine. Courtesy of Union Carbide. (c) Proof of the uniformity of a parylene coating. Note that the coating on the shaft and over the point is the same thickness. Photo courtesy of Dr. Wm. Beach, Bridgewater, New Jersey.

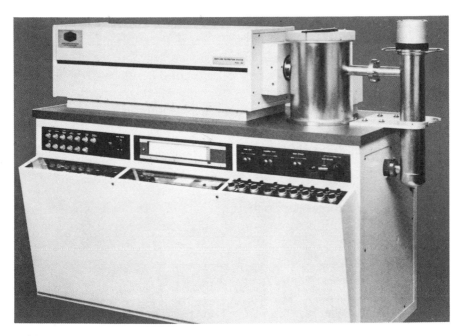

(b)

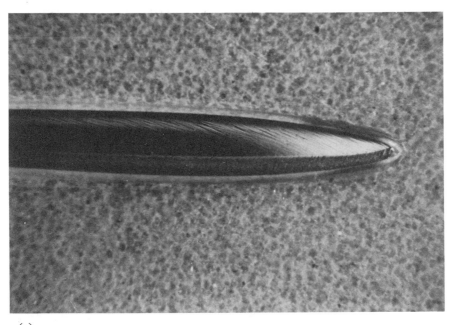

(c)

the reactive vapors of paraxylylene. This vapor enters the deposition chamber at room temperature, where it simultaneously deposits and polymerizes on the substrate.

It has penetrated every crevice and hole. Parylene does not, however, bridge holes. If portions of components or boards should not be coated, those areas should be protected by masking.

B.   Advantages and Disadvantages

Advantages

Forms a pinhole-free coating.
Has a very uniform coating thickness, even over edges and points.
Does not bridge holes.
Coated objects see only room temperature.
There is no runoff or sag.
The high vacuum pulls off all moisture prior to coating.

Disadvantages

Very capital intensive.
Raw material is very expensive.
Very difficult to remove for repair.
Cannot be soldered through (see Section IV.E of Chapter 7 for a discussion of a new, removable Parylene).

REFERENCES

1. Impregnation study of porous elements, *IEEE Trans. Electr. Insul.* Sept. 1985.
2. Pressure Gelation Process–SPI, Reinforced Plastics/Composites Institute. 29th Ann. Tech. Conf. 1974.
3. *Plastics Technol.*, May 1983, pp. 53–57.
4. *Plastics Technol.*, Sept. 1984.
5. *Thermoset Molding,* Society of the Plastics Industry, New York, 1970.
6. Overview of parylene coating methods, *Electronics*, Sept. 1985.

APPENDIX: SOURCES FOR MATERIALS

Mold Releases

Axel Plastics Research Lab., Woodside, New York
Camue-Campbell, Inc., St. Louis, Missouri
Ellen Products, Stony Point, New York
Freeman Manufacturing and Supply, Cleveland, Ohio
Frekote, Boca Raton, Florida
Hysol Corp., Olean, New York
Miller-Stephenson Chemical, Danbury, Connecticut
Perry Harms Corp., Wheeling, Illinois
Sprayon Products, Bedford Heights, Ohio
Staken Corp., Encino, California
Techform Labs, Los Angeles, California
Tech Spray, Inc., Amarillo, Texas

Potting Shells

Hammond Manufacturing Co., Guelph, Ontario, Canada
Hudson Tool & Die, Newark, New Jersey
Micro Plastics, Chatsworth, California
Moldronics, Downers Grove, Illinois
Morris Enterprises, Sepulveda, California
Orion Resdel, Inc., Rio Grande, New Jersey
Plasmetex Closures, San Marcos, California
Plastronic Engineering Co., Haverhill, Massachusetts
Protective Closures, Buffalo, New York
Starnetics, Van Nuys, California
Stevens Products, Inc., East Orange, New Jersey

Liquid Injection Systems/Meter Mix Dispense

Accumetric, Inc., Elizabethtown, Kentucky
Advantek, Inc., Minneapolis, Minnesota
Ashby Cross, Danvers, Massachusetts

Automatic Unlimited, Inc., Woburn, Massachusetts
Fishman Philip Corp., Holliston, Massachusetts
Glenmarc Manufacturing, Northbrook, Illinois
Hull Corp., Hatboro, Pennsylvania
Isochem Resins, Lincoln, Rhode Island
Kahnetrics Dispensing System, Riverside, Connecticut
Liquid Control Corp., North Canton, Ohio
Methods Engineering, Vauxhall, New Jersey
Midland Ross Co., Cambridge, Massachusetts
Otto Engineering, Carpentersville, Illinois
Pyles Corp., Wixon, Michigan
Richley Products, Fremont, California
Sealant Equipment Engineering, Oak Park, Michigan
Tridak Division—Indicon, Brookfield, Connecticut
Twin Rivers Engineering, East Boothbay, Maine

Compression Molding Equipment

. Dake Corp., Grand Haven, Michigan
Greenerd Press & Machine Co., Nashua, New Hampshire
Hull Corp., Hatboro, Pennsylvania
Kras Corp., Fairless Hills, Pennsylvania
P.H.I., Industry, California
Stokes Division, Pennwalt, Philadelphia, Pennsylvania
Technical Machine Products, Cleveland, Ohio
Wabash Metal Products, Wabash, Indiana

Injection Molding Equipment

Autojector, Inc., Albion, Indiana
Dake Corp., Grand Haven, Michigan
Gluco, Inc., Pittsburgh, Pennsylvania
Hull Corp., Hatboro, Pennsylvania
Illinois Precision Corp., Wheaton, Illinois
Jaco Manufacturing Co., Berea, Ohio
Morgan Industries, Long Beach, California
Stokes Division, Pennwalt, Philadelphia, Pennsylvania

Transfer Molding Equipment

Dake Corp., Grand Haven, Michigan
Gluco, Inc., Pittsburgh, Pennsylvania
Hull Corp., Hatboro, Pennsylvania
Kras Corp., Fairless Hills, Pennsylvania
Lauffer Pressen, Horb, West Germany
Micro Robotics, Lowell, Massachusetts
Pasadena Hydraulics, Industry, California
Stokes Division, Pennwalt, Philadelphia, Pennsylvania
Technical Machine Products, Cleveland, Ohio
Wabash Metal Products, Wabash, Illinois

Conformal Coating Application Equipment

Conforming Matrix, Toledo, Ohio
Glenro, Inc., Paterson, New Jersey
Humiseal Division Columbia Chase, Woodside, New York
Hyestis Machine Corp., Bristol, Rhode Island
Integrated Technologies, Acushnet, Massachusetts
Union Carbide (Parylene only), Danbury, Connecticut

5
Material Selection

I. STEPS IN SELECTING AN ENCAPSULANT

The encapsulant used to protect an electronic device is seldom the ideal or even the best available. It is usually selected because it will give satisfactory performance at an acceptable cost and can be applied by a method appropriate to the equipment and personnel available in the facility. For instance, the encapsulant finally selected will not necessarily be the one with the best electrical performance. Other demands are made beyond electrical performance in the device—demands such as:

Is a major capital investment necessary to permit its use?
Example: Transfer molding.
Can it be handled by the type of personnel available to apply it?

Is it cost-effective?

Must a defective device be repaired after encapsulation, or is it a throwaway? If the answere here is "reclaim," this is *a* major, if not *the* major factor in selection of a material.

Do not overengineer when material selecting. Determine the minimum protection that will satisfy the performance demands and seek a material that will provide this at the best economics. The bigger the company, the more remote are the people writing the material specifications from the economics and practicalities of the device. "Keep it simple" is not an original phrase, but it expresses a wisdom applicable to material selection.

A. Step 1: Pot, Mold, or Coat?

The first step in material selection is really a combination of material and method considerations. If a PC board requires a thin, transparent, environmentally protective coating, the choices are between the conformal coatings of class III. The method then becomes the one appropriate to the application of the material selected.

If, however, a coil is to be molded and cannot be in a can or shell, the first question is: How many are to be produced? If the answer is in the several thousands, transfer or injection molding as a method would be an early candidate. Other questions then deal with the selling price of the coil and its ability to support the high capital cost of the press and molds. If this is positive, the selection is from class II materials.

If the device is more complicated in design and geometry (than a simple coil) and cannot withstand the pressure or the capital costs of transfer or injection molding, a class I liquid will probably be required. A method will then be selected based on the number to be produced and the personnel and the facilities available and affordable.

The device to be protected may be a PC "card" needing protection only from airborne contaminates at ambient temperature.

Here a conformal coating may be adequate. In actual fact, the use of liquid encapsulants of class I satisfies the great majority of encapsulation needs, probably 80% or greater.

B. Step 2: Material Selection Begins

With step 1 completed, actual material selection can begin, but not, however, in absence of method. One influences the other too much for either one of them to be selected in absence of considerations of the other.

At this point it is essential that a full discussion with a material supplier be held. Be generous with background and end-use information, as this will greatly shorten the final material selection process. Too often a material is selected and even samples evaluated based on incomplete information forthcoming from the user, only to be changed. Why? "Oh, I forgot to tell you that!"

Make a realistic list of performance parameters of the device. List your facility and personnel "haves" and "have nots" and "cans" and "cannots" and add in the "wants." Example: I won't buy a dispenser. The few moments or hour or two spent on this will save hours, money, and aggravation later.

II. SELECTING THE MATERIALS

A. Class I Materials

1. More Questions

More questions must now be asked and answered. We must decide whether an epoxy, RTV silicone, polyurethane, or polyester is necessary or will do the job. "Is necessary" and "will do the job" are important phrases; the difference between "an RTV silicone is necessary" and "a low-cost epoxy will do the job" is a lot of money. There is no simple method or answer here. There are just too many considerations intertwining method, material, performance, economics, facilities, personnel, and so on.

Pertinent Information Required

What performance is demanded of the device and therefore
    of the encapsulant?
What cost can be tolerated for the material? Be honest now;
    it is not a "wish list" you are making.
What application equipment is available, or is it necessary to
    apply the material by the simplest method—hand-mix
    and pour?
Does the device contain stress-sensitive components?
Is there a specification for mechanical shock requirements?
What cost of material per device can be tolerated? Be realistic.
    Do not expect silicone performance if the maximum
    cost per cubic inch that can be tolerated is $0.10.
Must postencapsulation repairs be made, or can a defective
    device be thrown away? The answer to this question is
    critical. If it is "yes," a defective encapsulated device
    must be reopened for rework and repair on a regular
    production basis. Then a silicone encapsulant is almost
    dictated, as it is the only material that permits this on a
    production basis. Soft epoxies and urethanes, although
    too tough and adherent for most repair situations, may
    permit repairs under specialized circumstances only.
If a liquid material is to be hand-mixed and poured, a long
    pot-life system would be appropriate. But remember, a
    long pot life means slow cure. Is this acceptable? Speed-
    ing the cure with heat means equipment and energy,
    which means many dollars of investment.
Is a high thermal conductivity necessary? If so, it will affect
    the next question.
If a meter-mix-dispenser is to be used to minimize labor and
    volume of throughput, should it be of fixed or variable
    ratio? If other potting jobs are done or are to be done, a
    variable-ratio machine may be cheaper. Also, the type of
    filler in the encapsulant is now important, as it can
    seriously abrade dispenser components. This is particu-
    larly true of high-thermal-conductivity systems.

Is a room-temperature cure necessary because of plant and
equipment limitations?

If the device is a high voltage one, is it also high frequency?
If so, silicone may be the only encapsulant that will perform in the device.

Is a fast-room-temperature cure of a large mass of material
necessary?

Does the device contain heat-sensitive components? This may
preclude heat-cure or fast room-temperature-cure
epoxies, urethanes, or polyesters, as they generate considerable heat. Is the encapsulant functional or mainly
cosmetic with only minor functional demands? Must it,
in addition to the electrical performance demands, resist snoopers? If so, encapsulants that are heat cured and
contain abrasive fillers are necessary.

What thermal cycling requirements are there? Silicones offer
the widest range. But if the cycle is low to medium high,
a urethane may do. If it is moderately low to high, an
epoxy may.

The answers to such questions as these *and more* will steer
the user, first, to a method and material type, then to an individual formulation of that material type.

Mechanical and Facility Considerations. Now that you have
selected and are about to decide on an encapsulation material,
some other considerations are important.

Can the first-choice material be handled in the existing plant
facilities and equipment? Will capital expenditure be required? If the first-choice material requires a heat cure
and no ovens or capital for them is available, on to
choice number two.

The capability of available personnel to apply the material is
a very important consideration. Can you dedicate one
person to do the potting? If more than one person attempts to do it, the potting procedure *will be a disaster*.
If a dispenser is involved and anyone can "have a go at

it," an even bigger disaster will result. One person should be assigned as the "father confessor" to meter-mix-dispenser equipment. Too many cooks starting up and shutting down a dispenser will prove disastrous to the broth.

One-to-one ratio systems are easier to handle successfully than, say, a 6 or 7% ratio, both through a dispenser and when hand-mixing.

Personnel differ radically in their reaction to certain chemicals used in some epoxy and urethane systems. Who will handle these?

2. Epoxies

Composition. A class I epoxy formulation will have the following general composition:

Liquid epoxy resin
Curing agent
Mineral filler (see below)
Viscosity reducer (optional)
Color pigment (optional)
Additives: accelerators, defoamers, and so on

Mineral fillers are not used in very low viscosity impregnating materials. The filler particles are filtered out by all but the coarsest windings and form a "filter cake," which effectively inhibits further impregnation.

Property Profile. Liquid epoxy systems, being the largest class of materials used in encapsulation, are usually the first materials to be considered. The various epoxy formulations available offer a very wide profile of properties:

Very soft systems available
Very hard systems available
High-thermal-conductivity systems available
Good thermal insulation systems available
Excellent dielectric strength

Excellent electrical conductivity systems available
Long pot life
Very fast cure
Some curing agents cause sensitivity problems
Excellent adhesion to most electronic materials and substrates
Good resistance to aggressive chemicals
Medium-high dielectric constant and dissipation factor
Fair low-temperature strength and shock
Good high-temperature strength and resistance

Use Considerations

Fast-cure systems generate considerable heat. This becomes an increasing problem, as the mass of the casting increases since the heat is generated faster than it is conducted away and can build up to destructive proportions.

Epoxies become hard and brittle at temperatures of -10°C and below. When considering epoxies for devices cycling down to these temperatures, this must be considered.

Very flexible systems, such as the epoxy gels, have lower insulation resistance and higher dielectric constant and dissipation factor than RTV silicones and more rigid epoxies.

Many systems can be made in a 1-to-1 ratio, eliminating a major source of error—measuring the curing agent.

Very low cost systems are available.

Epoxies generally have higher dielectric constant and dissipation factor than silicones and urethanes.

Although soft epoxy gels are repairable, they are not nearly as easily repairable as silicones.

Although any encapsulating system can be mishandled, epoxies are among the easier to use. Nothing inhibits their cure.

One-component (no curing agent measurement errors) systems are available, but must be heat cured.

Some epoxy curing agents can cause dermatitis and respiratory problems in certain individuals. Although good hygiene and ventilation are important in handling any of
these systems, they are particularly important with
room-temperature-cure epoxies.

Much lower cost than silicones and about the same general
price range as urethanes.

Optically clear systems are available.

Widest range of formulations available of the class I systems.
For instance, all available mineral fillers are used in
epoxy systems. This means that high- and low-thermal-
conductivity systems are available as well as electrically
conductive, easily machined, and "secure" (i.e., very
hard to enter) systems. Very thick dipping and thin
impregnating epoxies are used.

3. Silicones

Composition. An RTV silicone compound will have the following general composition:

Base silicone polymer
Cross-linker
Catalyst
Mineral filler*
Color pigments (optional)
Viscosity reducer (optional)

Property Profile. Many encapsulation jobs almost dictate
the use of silicones, such as where the cost of the constituent components make it necessary to remove and replace components as
opposed to junking the whole device. Silicone's low adhesion or

---

*Mineral fillers are not used in clear silicone formulations, either to keep
viscosity low or, more important in silicones, to maintain their crystal clarity,
desirable when doing component replacement. Fillers *are* important, even essential, in all other silicones, and they make a major contribution to the development of physical properties, which would otherwise be very low. This is
in sharp contrast to filler use in epoxies and urethanes.

no adhesion (without a primer), easy "cutability," low tear strength, and low tensile strength make it very easily removed for such repairs. Upon completion of the repair, silicone reseals to itself perfectly. RTV silicones stand alone in their ease of repair.

Some silicones revert
No hard or brittle systems
Poor solvent and chemical resistance
No dermatitis potential
Best low-temperature flexibility
Best electrical properties
Best high-temperature electrical properties
Very low tear and physical strength
Very high coefficient of thermal expansion
Highest cost of the class I materials
Some can inhibit and not cure completely
No exotherm on cure
Can be cast in infinite thickness
Best repairability
Widest-performance temperature range; the better silicones
    offer a range of - 100 to +250°C

Use Considerations

Most easily repaired of all the encapsulation materials. Can be
    cut away, a repair effected, then resealed with fresh ma-
    terial very easily.
There are no rigid silicones. If the device is to be cast and the
    encapsulant is to be the wearing surface, silicones will
    probably not be satisfactory.
Have a very high thermal coefficient of expansion. The
    highest of any of the class I materials. Even though soft,
    silicones can break things if not given room to expand
    during thermal excursions. This property varies widely
    among the available silicones.
Best and lowest low-temperature flexibility—to - 55°C—with
    special silicones offering a brittle point to - 115°C.

Best high-temperature electrical properties.

No exotherm during cure; therefore, no destructive heat
buildup in large masses.

No adhesion without use of primers*; they self-adhere, however.

Highest cost of the class I materials: two to five times that of
epoxies and urethanes.

Lowest dielectric constant and dissipation factor.

Highest dielectric strength and insulation resistance.

Some silicones can be cure-inhibited when in contact with
certain materials. The addition-cure silicones are the offenders here. Materials that can cause problems are heat-
cured rubbers, amine-cured epoxies, soldering flux residues, vinyl wire insulation, and others.

Some silicones can "revert" upon exposure to very high
temperatures (400°F) or high humidity and high
temperature. The condensation-cure systems offend
here.

Physical and electrical properties change little over a wide
operating temperature range—much less than do epoxies
and urethanes.

One-component heat-cure silicone potting and casting systems are not available as yet.

Special note. One-component moisture-cure silicone sealants
do not have utility here, as they require exposure to atmospheric
humidity to cure. The humidity releases the curing agent
"blocker," which allows the cure to proceed. A material (the
cross-linker block) is released during cure and must escape for cure
to proceed. Practically, these one-component sealants will cure to
about ¼ in. depth before the lack of moisture, and the inability of
the effluent material to escape, stops the cure. They are useful,
however, as sealants and adhesives, to hold wires down, com-

---

*Two-component RTV silicones, which adhere without the use of primers,
are becoming available as of early 1985. Cost and inconsistent performance to
date are slowing their growth, however.

ponents "on," and so on, as they do adhere tenaciously to most substrates.

4. Urethanes

Composition.   A polyurethane compound will have the following general composition:

Isocyanate prepolymer
Curing agent
Viscosity reducer (optional)
Mineral filler (optional, used infrequently)
Color pigment (optional)
Catalyst

Property Profile.   Urethane systems offer a faster cure with generally lower exotherm than epoxies. However, their fast cure results in a very short working life, often too short to permit a vacuum deairing step. But since most urethanes do not contain mineral fillers, vacuuming is often not necessary or can be done very briefly, as there is no filler to slow the process. But (again) no filler means higher shrinkage on cure. However, there is one other major consideration when considering urethanes, which is their extreme moisture sensitivity in the wet state. They must be isolated from contact with atmospheric moisture (humidity) at all times. This is a major consideration to be taken into account when selecting an encapsulant.

Good low-temperature flexibility
Fast cure
Moderate exotherm
Poor high-temperature physical and electrical properties
Uncured material is extremely moisture sensitive
Short pot life
No one-component systems
Can revert under some conditions
Few high-thermal-conductivity systems
Not easily repairable

Higher shrinkage than epoxies and silicones

The vapors of the isocyanate-containing portion are noxious
and should not be breathed

Use Considerations

Offer lower brittle points than epoxies. Cycle easier to lower
temperatures than epoxies but not as low as silicones.
Average range to -40 or -50°C.

Do not retain their physical and electrical properties to as
high a temperature as epoxies. Average range to +125°C.

The vapors of the isocyanate portion in most urethane systems can be very noxious and should not be breathed. It
is for this reason, more than any other, that the basic diisocyanate is never present as such but only as a prepolymer. Because of the potential for residual isocyanate, do not breathe the vapors and use only with
good ventilation.

Urethanes cure very quickly with a lower exotherm than that
of epoxies. However, in large masses, this exotherm can
still be considerable and is still a potential problem due
to thermal expansion, then shrinkage, with subsequent
stress.

Because of their fast cure, urethanes have an extremely short
pot life, making handling and degassing difficult in
hand-poured systems.

Uncured material is very sensitive to any moisture—such as
humidity—and must be protected. Cans should be immediately reclosed and it is generally recommended by
suppliers that they be backflushed with dry nitrogen
after each use.

Lower dielectric constant and dissipation factor than epoxies,
but not as low as silicones generally.

Soft urethanes are not easily repairable because of their
toughness and adhesion. No better in this respect than
soft epoxy gels.

Solvent and chemical resistance is poorer than epoxies but

are commensurately easier to strip away where this is desirable. Better than silicones in resistance.

Conditions of high humidity (60%+) and a temperature of 100°F and higher often cause urethanes to revert, actually returning to a liquid state. Where this is critical, discuss thoroughly with your material supplier. The Naval Avionics Laboratory has developed procedures for testing for reversion.*

Prices of urethane compounds are in the same general range as epoxies and, of course, some one-half to one-fourth that of silicones.

Urethane systems seldom contain high loadings of mineral fillers due to their extreme moisture sensitivity and the difficulty of drying fillers to a sufficiently low moisture content. Consequently, the range of properties available in urethanes is much narrower than in epoxies.

3. Polyesters

Composition

Base polyester resin
Reactive diluent, usually styrene
Mineral filler, often sand
Accelerator
Catalyst

Property Profile. Polyesters are the cheapest but offer the poorest properties of the four liquid systems.

Fast cure
High exotherm
High shrinkage
Pervasive odor
Low viscosity
Flammable in uncured state

---

*Naval Avionics Center, Indianapolis, IN 46218.

Limited shelf life
Flexible systems available
Very poor solvent resistance
Poor high-temperature properties

Use Considerations

A very prominent and pervasive odor from styrene monomer,
an integral part of the polyester system. This cannot be
eliminated and is much stronger than that from any of
the other three liquid systems.
Very flammable in the liquid state due to the above-mentioned
styrene diluent. Flames and sparks must be eliminated
from the area.
Very fast cure depending on amount and type of accelerators
and activators added and the amount of peroxide
catalyst used. Polyesters are faster than epoxies, although
urethanes can be formulated to be comparably fast and
with lower exotherm.
Very high exotherm, comparable to that of fast-cure epoxies.
Poorest high-temperature physical and electrical properties of
the four liquid systems. Do not design for greater than
175°F continuous operating temperature.
Low viscosity. Adding additional styrene does not seriously
degrade properties and greatly improves penetration in
sand potting.
Difficult to use through a dispenser due to low level of
catalyst used, often less than 1%.
Most solvent sensitive of the four liquid systems.
Polyester resin is very sensitive to storage temperature and
can begin spontaneous polymerization (cure) if stored in
hot areas despite containing a cure inhibitor.
Polyester resin is used rarely because of the many drawbacks
listed previously. It is used mainly where cost outweighs
all other considerations and is usually "formulated" by
the end user. A drum of general-purpose "boat"-grade
resin is purchased, mixed with a filler (usually sand),

and the device potted. Conversely, the unit is packed with sand and the polyester is allowed to trickle down and through.

6. Fillers in Liquid Encapsulants

Once the type of potting/casting/dipping/impregnating material has been selected, other considerations need to be taken into account. The first question that should be asked is: Should it be a filled or unfilled system? If circumstances dictate that the absolutely lowest viscosity is essential, such as would be the case if thoroughly impregnating a coil, then an unfilled system would be best, and it is hoped by now the obvious choice, since any filler particles will raise the viscosity significantly and probably be filtered out by the windings. They would form a filter cake and act as a barrier to further impregnation [1]. However, such circumstances as these are rare, as by far the greatest number of encapsulation jobs are in potting or casting a unit and a filled material is desirable and is used.

*Effects of Fillers in Liquid Encapsulants.** Why add fillers to encapsulants? Because they contribute many desirable properties, such as:

Decrease cost.
Decrease shrinkage.
Lower the peak exotherm in epoxy urethane and polyester systems. They do this by absorbing part of the heat generated during cure.
Improve mechanical strength, particularly important in silicone encapsulants.
Improve dielectric strength and insulation resistance.
Lower the coefficient of thermal expansion.
Improve thermal shock.
Raise thermal conductivity.

---

*For more information, see Ref. 2.

The right choice of filler can make encapsulants either electrically conductive or thermal insulators.
Increase pot life.

and a few undesirable ones, too:

Raise dielectric constant and dissipation factor.
Often settle out of suspension and must be reincorporated
by stirring or shaking before use.
Retain air bubbles: both cosmetically and functionally unde-
sirable.
Increase viscosity.
Fillers are often abrasive and can cause rapid wear in handling
and dispensing equipment.

Filler Types. The most commonly used fillers are:

Silica: three different types
Calcium carbonate: natural and precipitated
Fumed silica, often called Cabosil*
Aluminum oxide: two types
Mica
Microballoons
Silver and gold
Other

Silica and calcium carbonate. These two are the least expensive and therefore the most widely used, probably accounting for
more than 75% of all filler used in liquid encapsulants. They are
used in the form of very fine—325 mesh—powders and are incorporated by the formulator into the liquid resin. Both are mined
from natural deposits and ground to powders. They are used for
the most basic of reasons:

They lower the cure exotherm.

---

*Trademark of Cabot Corp.

They decrease shrinkage.
They lower cost.
They improve the mechanical properties.

Differences between the two.   *Silica* is very abrasive, making machining and tapping difficult with regular tools, and has a deleterious effect on dispensing equipment. Raises the viscosity of the liquid resin slightly more than calcium carbonate. *Calcium carbonate* machines and therefore taps very easily. Has slightly poorer moisture resistance than its companion and gives a lower-viscosity formulation.

Three types of silica

1. *Microcrystalline*: most widely used. It has the least effect on viscosity of the three types and readily releases air bubbles. Settles out of suspension readily. It is mined from natural deposits and is the cheapest.
2. *Cryptocrystalline*: often and erroneously called amorphous silica. A much smaller particle size than microcrystalline and has a greater effect on viscosity. Retains air bubbles quite tenaciously, but is quite resistant to settling. It is also mined from natural deposits and is slightly more expensive.
3. *Amorphous*: often called fused silica. It is a naturally occurring silica that has been reprocessed through a fusion stage. It is lower density than microcrystalline and displays a lower coefficient of thermal expansion, desirable for compounds that must undergo thermal cycling. Three to five times more expensive than the previous two.

Two types of calcium carbonate

1. *Wet ground and washed*: the more common of the two types. Gives the least viscosity rise when dispersed and releases air bubbles quite readily. Settles less readily than microcrystalline silica and is easier to reincorporate. It costs about the same as microcrystalline silica.

2. *Precipitated*: natural material that has been dissolved, then precipitated. The dissolution step has two benefits: (a) easy removal of impurities, and (b) controlled particle size on reprecipitation. The controlled particle size allows better batch-to-batch viscosity control. It costs about 50% more than wet ground material.

Aluminum oxide. Available in two distinctly different forms, aluminum oxide is a truly functional filler, used only where its unique properties are required.

Tabular alumina. Highly abrasive, very dense, and has the least effect on viscosity of any of the major fillers. But its real claim to fame is high thermal conductivity, which combined with its minimal effect on viscosity permits incorporation of large amounts into the liquid resin, resulting in encapsulants that are efficient thermal conductors.

Tabular is also some five times as expensive as silica or calcium carbonate, so is used for its functional properties. A second functional property sometimes used to advantage is its extreme abrasiveness. This can make circuitry quite inaccessible to spies or snoopers. This, combined with the right resin hardener combination, can make chemical stripping difficult also.

Comparative thermal conductivities of silica, calcium carbonate, and tabular alumina for "average" potting materials are:

Silica: 4–6 Btu/ft² /in./°C/hr.
Calcium carbonate: 3–5 Btu/ft² /in./°C/hr.
Tabular alumina: 9–12 Btu/ft² /in./°C/hr.

Hydrated alumina. This other form of aluminum oxide is also known as aluminum trihydrate. It contains water molecules "locked" into the crystalline structure of the particles. Locked up this way, the water is not troublesome to the base resin in the formulation. However, when the temperature reaches approximately 650°F, it is spontaneously released and acts to snuff out flames and suppress noxious fumes in any burning plastics that contain it. Hydrated offers an inexpensive route to Underwriters' Laboratories (UL) grade 94 encapsulants. Costing some two to three times as much as silica, it, too, is used where this functional property is necessary. It is not as abrasive as tabular but raises viscosity more.

Mica. Mica is another functional filler. It exists as platelike particles, easily broken during processing. It acts as a reinforcement where thermal cycling properties are important, particularly in dip-coated devices. It is more expensive than silica and causes a much higher, even beneficial viscosity increase in dip encapsulants. Care must be taken during its incorporation into the encapsulant to avoid excessive breakage of these particles.

Microballoons. As the name implies, this filler is composed of tiny hollow spheres usually made of glass, but plastic is also available. They are used for two reasons:

1. *Flotation*. Microballoons' very low density permits encapsulants to be formulated with specific gravities below 1.0 either for low weight, where this is important, or for actual flotation devices.
2. *Low dielectric constant*. Because they displace a lot of relatively high dielectric constant resin, with a thin-wall glass envelope of air, microballoons significantly lower the dielectric constant of resin systems in which they are incorporated.

Microballoons cause several other phenomena in encapsulants:

1. Even more easily broken than mica platelets, so extreme care must be taken during the dispersion stage.
2. Significantly lower the dielectric strength of formulations. This is probably due to their smooth exterior wall, resulting in less mechanical locking of the resin to it during cure.
3. They float. Microballoons are unique among fillers in that they do not settle on storage but rather, float. They must therefore be "stirred down" before the encapsulant is used.

Silver and gold. In the form of tiny flakes, gold and silver are used to impart electrical conductivity to normally dielectric thermosetting polymers. They find use in "cold" solders and in radio-frequency and microwave attenuating coatings, sealants, and gaskets. Such formulations are obviously very expensive, being priced in accordance with the fluctuating world market price for precious metals.

Other fillers.   Several other fillers may be encountered occasionally and merit a brief mention.

Sand.   This is the coarse, water-washed version of the 325-mesh microcrystalline silica encountered earlier. It is used as a packed-in dry filler, which is subsequently impregnated with a low-viscosity unfilled encapsulant. It occupies space inexpensively and removes heat. This technique is often called "sand potting."

Wollastonite.   A somewhat fibrous naturally occurring mineral. It reinforces against mechanical and thermal shock. It is used as a thickener and reinforcement in dip encapsulants.

Barytes.   A barium sulfate-based mineral—very dense. It has a minimal effect on viscosity, so a high loading can be achieved, but unfortunately, it settles severely. It is sometimes used to shield circuitry from x-ray snooping.

Clays.   Build a nice, light thixotropy in adhesives and dipping materials. Because of their natural tendency to be like ion exchange resins, they are difficult to make "electronically" clean.

Talc.   A light, powdery mineral, it imparts a puffy-type flow control in adhesives and dipping materials.

Cabosil.   Also called "fumed silica," a description that comes from its method of manufacture. It is a light, fluffy form of silica, resembling household dust in appearance. Small amounts (1 to 5%) can turn a free-flowing liquid encapsulant into a heavy paste. Smaller amounts (¼ to 1%) act as a filler-settling retarder. When surface-treated with a silicone coupling agent, this form of silica is used as a toughner in silicone rubbers.

Filler Selection.   The type of filler in the encapsulant that you have selected becomes of significance only if you need one of the functional properties offered by a particular filler or if you are going to use metering equipment. Tabular alumina is so abrasive that mechanical dispensing becomes difficult to the point that it becomes prohibitive. Its use necessitates continual part replacement as they are abraded away. Silica is less abrasive but is still a problem for dispensing equipment. So if a dispenser is to be used, select a calcium carbonate–filled encapsulant. It will have a minimal effect on the equipment, permitting much longer service life, but it still must satisfy all other performance criteria.

The filler type to be used in your encapsulant of choice should be discussed with an encapsulant supplier. Explain your needs and special problems. Do not have preconceived notions about materials. Let the formulator do the formulating; you merely present the performance parameters.

Following is a filler summary.

Filler	Major functional property	Relative cost
Silica	Low cost	Lowest
Calcium Carbonate	Low cost, machineable	Very low
Tabular Alumina	High thermal conductivity	Medium
Hydrated alumina	Self-extinguishing	Medium low
Gold/Silver	Electrically conductive	Highest
Cal-o-sil	Thickener	High
Mica	Reinforcent	Medium
Wollastonite	Reinforcent	Medium–low
Talc	Puffiness—for dipping and adhesive materials	Medium–low

7. Summary of Class I Materials

The vast majority of potting jobs are satisfied by one of the thousands of epoxy systems in use. From large castings (hundreds of pounds) to small, fast-cure "blobs" on a PC board, epoxies do most electrical and electronic protection jobs well. Where high-reliability military or aerospace application and/or very high voltage is the end use, RTV silicones are often selected regardless of the price penalty. For such uses repairability is often an over-riding consideration. When fast production demands a fast cure and where low-temperature flexibility is important, a urethane is often the choice, despite moisture and handling considerations.

Polyester is used only where cost is the one overriding consideration, outweighing its drawbacks of odor and flammability.

Keep in mind that when starting with a rigid system,

As flexibility increases, moisture resistance deteriorates (except silicones).
As flexibility increases, thermal coefficient of expansion increases.
As flexibility increases, insulation resistance decays (except
silicones).
As flexibility increases, dielectric strength decreases (except
silicones).

Also keep in mind that when starting with a clear, unfilled system:

As the filler content increases, the exotherm on cure de-
creases.
As the filler content increases, shrinkage on cure decreases.
As the filler content increases, thermal coefficient of expan-
sion decreases.
As the filler content increases, cost decreases.

Also, in a filled system, as the filler level increases:

Pot life increases.
Cure time increases (cures more slowly).
Thermal conductivity increases.
Specific gravity increases.
Viscosity increases.
Air retention increases.

Following is a comparison table of the properties of various
class I materials.

Property	Silicone	Epoxy	Urethane	Polyester
Operating tempera- ture range	Widest	Low to high	Very low to medium low	Smallest

Property	Silicone	Epoxy	Urethane	Polyester
Brittle point	Lowest	Medium	Low	Poorest
Relative cost	Highest	Medium to low	Medium to low	Lowest
Shrinkage on cure	Lowest	Low	Medium to low	Highest
Adhesion	None (without primer)	Best	Good	Fair
Coefficient of thermal expansion	Highest	Lowest	Good	Fair
Repairability	Best–very good	Poor to fair	Poor to fair	Poor
Mechanical strength	Lowest	Best	Good	Good
Pot life	Short to medium	Short to long	Short	Short
Exotherm on cure	None	High	Medium to high	Highest
Flammability –				
wet	Very, very low	Very low	Very low	IIigh
cured	Very, very low	Very low	Very low	Fair
Ease of use	Good	Easiest	Good	Most difficult
One component	No	Yes	No	No
Shelf life	Medium to long	Longest	Medium	Shortest

*Safe Handling.** These materials must be treated as what they are—reactive chemicals. Use eye and hand protection and good ventilation. Store in cool, dry places. Avoid skin contact and inhalation of vapors.

Silicones present a minimum of handling hazards. The catalysts of the condensation-cure systems can cause eye irritation.

Epoxies are fairly safe, though some people react to some diluents in them. As it is impossible to predict who will react, practice prevention. If contact is made, wash as described previously or use the hand cleaner. Do not use solvents: this greatly aggravates any problems. Do not breathe any vapors. Ventilate work areas.

---

*For more information, see Ref. 3.

Epoxy Curing Agents are the heavies in this drama. They are both very reactive chemically and hyper-allergenic. They can cause chemical burns and allergies (similar to poison ivy and asthmatic-like responses) in some people. Make maximum efforts to prevent any skin contact or breathing of vapors. Even if you haven't had a problem, you might, as some people tolerate them for a time, then react (even after a year or more) and subsequently become hyper-sensitive. Ventilate work areas. If skin contact occurs, wash area immediately. If it gets in the eyes, flush with lots of water and call a doctor. If ingested, call a doctor immediately. If clothing has been contaminated with any of these compounds or catalysts, do *not* wear it until it has been thoroughly laundered.

Urethanes — the base material is rather like an epoxy as far as skin contact is concerned, but the vapors should be avoided as they can be very noxious.

Urethane Curing Agents should be regarded and handled like epoxy curing agents as they often contain amines and can cause both allergenic responses and respiratory problems.

Polyesters should be handled as you would solvent-containing coatings, keeping them away from flames, sparks, etc.

Polyester Peroxide Catalysts are strong (oxidizing) agents and should be handled accordingly.

B. Class II Materials

When selecting a material for transfer, injection, or compression molding, there are only a small fraction of the choices, faced with liquid systems. Not only are there many fewer suppliers, there are also fewer choices per supplier. There is much less variety in property variations of molding powders than in liquid systems and no degree of freedom in curing agent variations is permitted the user. Usually, the component to be molded and its end use will further shorten the already short list of contending materials. Then the method of encapsulation and the type of part will dictate the type of material. The choice is then among a very short list of candidate materials.

1. Epoxies

Composition. An epoxy powder will have the following general composition:

Base resin (Bis A)
Curing agent
Accelerator
Mineral filler
Release agent
Filler reinforcement (optional)
Catalyst
Color pigment (optional)

These ingredients are fluxed together at a temperature above the softening point but below the cure temperature until uniform, then ground and classified to the desired particle size.

Property Profile. Epoxy powders offer fast, easy flow and low molding pressure:

Fast flow and cure
Hard, tending to brittle
Fair to good impact strength
Medium shelf life; should be refrigerated
Good flexural strength
High-temperature classification: F/H
Expensive

Use Considerations. Fast, easy flow makes epoxies excel at encapsulating components such as coils, capacitors, semiconductors, and so on, in multicavity molds, where the long path the molten charge must traverse makes easy flow essential. The consequence of premature gelation is incomplete cavity fill and wasted material and components.

2. Phenolics

Composition. A phenolic powder will have the following general composition:

> Base phenolic resin: solid
> Curing agent: amine type
> Mineral fillers
> Fiber reinforcement (optional)
> Accelerator
> Color pigment (optional)

Just as with epoxies, the ingredients are fluxed together until uniform, then ground and classified.

Property Profiles. Phenolics are also "easy molders," but not quite as easy as epoxies. Most phenolics evolve ammonia during cure, and sometimes the mold must be opened to vent when compression molding.

> Easy molding, but higher pressure than epoxy
> Higher shrinkage than epoxy or DAP
> Very long shelf life; no refrigeration necessary
> Lower cost than epoxy or DAP

Use Considerations. The very high physical strengths and high-temperature rating and lower cost find phenolics used in "rugged" devices such as automobile distributor caps, coil caps, and consumer wiring devices.

3. DAP

Composition. A DAP molding powder has the following general composition:

> Base DAP resin

DAP monomer
Catalyst: solid peroxide
Mineral filler
Fiber reinforcement
Accelerator
Color pigment

These materials are dry-blended until uniform, then ground and classified. Fluxing can be and is occasionally used, but as with the epoxies, the temperature must be kept below the peroxide decomposition point.

Property Profile. The molding characteristics of DAPs are those of slow flow, high pressure, and poorer adhesion.

Very high physical and impact strength
Very high heat resistance
Long shelf life without refrigeration
Expensive
Fast molding cycles

Use Considerations. High molding pressures suggest DAP use as blocks and bases, subsequently fabricated into connectors, switches, and so on, rather than coil or resistor encapsulation.

4. Bulk Molding Compound: Polyester

Composition. A BMC has the following general composition:

Liquid polyester resin
Catalyst: solid peroxide
Glass fibers
Mineral fillers
Accelerator
Color pigment (optional)

These ingredients are wet-blended together in a dough mixer until uniform, then compacted and extruded into logs approximately 2 in. in diameter for shipping; always refrigerated.

Property Profile. Polyester BMC is the fastest molding of the class II materials, with the shortest shelf life.

Fastest molding
Least expensive
Lowest temperature limit
More difficult to use
High odor
Short shelf life

Use Considerations. BMC's short shelf life and handling disadvantages suggest its use in compression molding, as its stiff, doughy consistency makes transfer or injection impractical. It has low impact strength in thin-wall sections but finds use in large cases, housings, and switch bases. Odor and flash point are other drawbacks to its use. Cost is the driving force for its use.

5. Safe Handling

Powdered materials hold within themselves the potential of dust and dust hazards. Dust can be ignited by electrical sparks or open flames. All sources of dust should be contained in enclosures that are vented. A strong exhaust fan should also be used. A dust mask should be worn by operators near the dust generation point. All equipment should be grounded.

C. Class III Materials

As with class II materials, class III conformal coatings demand far fewer choices, both in terms of individual formulations available and methods of application. With Parylene, the material selection also dictates the method of application.

PC boards going into other than ideal end-use environments or ones that will not subsequently be potted as part of a larger device should have at least a minimum moisture and dust-resistant protective coating. This is most economically supplied by a conformal coating. Certain military aerospace boards, on the other hand, demand much higher-grade protection and, often, multiple coats of high-performance polymers are specified, even when the board will be subsequently part of a potted assembly. This is "going the extra mile" in the name of high reliability.

1. Acrylics*

Composition. An acrylic coating will have the following general composition, a very simple one:

Acrylic resin
Solvent
Additives (optional): dyes, pigments, and so on

These materials are simply mixed together until the resin has completely dissolved.

Property Profile. Acrylics offer very fast drying, with no pot- or shelf-life considerations and easy repair.

Easy to apply by dip, spray, or brush
Very fast dry (not cure)
No pot-life restrictions
Easily repaired; can be solvent removed
Low flash point, therefore dangerous; a few are available with
        "no flash" hologenated solvents
Good chemical resistance
Poor solvent resistance
Low solids
Very good short-term humidity resistance
Good long-term humidity resistance
Poor abrasion resistance
Temperature resistance low
High applied cost, as the solvent is a loss

Use Considerations. Since no resin cure is involved, only solvent evaporation, acrylics are the easiest to use, particularly for

---

*For more information, see Ref. 4.

short runs and prototyping. They are easily removed or touched up without loss of cosmetics. However, the price per gallon does not tell the whole story, as there is more solvent than resin and it is lost and must be added back into the cost per unit of dry film applied.

2. Urethanes

Composition. Polyurethane conformal coatings have a split "compositional" personality. There are both one- and two-component systems. Each would have the following general composition:

One-component:

> Isocyanate terminated polymer
> Solvent
> Catalyst
> Dyes, pigments (optional)

Two-component:

Part A	*Part B*
Isocyanate prepolymer	Cross-linker
Solvent (optional)	Solvent (optional)
	Catalyst
	Dyes, pigments (optional)

Property Profile. Moisture sensitivity in the wet state is the highest hurdle to successful urethane use.

> Easy to apply by brush, dip, or spray
> Fast dry to touch (solvent loss) but slow final cure
> Pot life: one-component—long, indefinite if protected from moisture; two-component—maximum of hours
> Heat acceleration of cure: one-component—no; two-component—yes
> Difficult to repair; selected "stripper" solvents are necessary
> Good solvent resistance

Fair chemical resistance

Humidity resistance: short-term—very good; long-term—very good

High, but not 100% solids. Less solvent loss than acrylics

Fair to good abrasion resistance

Low to medium temperature resistance

Low-flash-point solvents

Subject to reversion

Use Considerations. Both one- and two-component systems are used—the cure of one-components being activated by atmospheric moisture. The "cure" here has two stages, solvent evaporation and "chemical" cure. The former is completed quickly; the latter takes another 2 or 3 more days to reach ultimate properties. The board is handleable as soon as it is tack-free, the final cure proceeding without any further input from the user. Where maximum long-term moisture and humidity resistance is critical, urethanes make a good choice. The two-component systems cure much like epoxies with a similar pot life, viscosity buildup, gel, and hardening profile.

Both urethane types share a common problem of moisture sensitivity in the can, the one-component being the most sensitive. Most manufacturers recommend keeping lids tight, and even backflushing the cans with dry nitrogen after each use.

Polyurethanes are also considered hazardous because of residual free isocyanate in the vapors. Good ventilation should be provided and breathing vapors should be avoided.

3. Epoxies

Composition. Epoxy conformal coatings have a general composition not very different from a two-component urethane.

Part A	*Part B*
Epoxy resin, liquid or solid	Curing agent
Solvent	Solvent
	Dyes, pigments (optional)

Property Profile. With the exception of lack of moisture sensitivity, two-component epoxies handle like two-component urethanes.

Fast dry to handling, then slower final cure
Can be heat accelerated
Long pot life: hours, days
Difficult to repair; strippers strong enough to attack the
    epoxy can do likewise to the base board
Very good solvent resistance
Very good chemical resistance
Good short-term humidity resistance
Fair to good long-term humidity resistance
Flammable solvent, low flash point
Easy to apply by brush, spray, or dip
Very good abrasion resistance
High, but not usually 100% solids
Less solvent loss than acrylics and one-component ure-
    thanes
Medium temperature resistance

Use Considerations. All current epoxies are two-component systems. One-component systems are available but must be heat cured. The cure of epoxies is like that of the two-component urethanes with solvent loss (quickly), then slower final cure. When cured in high-humidity environments, epoxies sometimes exhibit the phenomenon known as "blush" or "spew." This is the reaction of any curing agent at the surface with atmospheric moisture and carbon dioxide. It shows up as a sticky, greasy feeling on the surface of the cured epoxy. It is most pronounced at lower temperatures and higher humidities. It can be mitigated by allowing the two components to induct for 15 to 30 min before coating. It is also much less pronounced in solvent thinned systems.

Epoxies are chosen where toughness, abrasion resistance and chemical resistance are important. They can be used to repair any reworked board, as they adhere well to all but silicone coatings.

Rules of good hygiene must be followed here just as with the epoxy encapsulants of class II.

4. Silicones

Composition. The general composition of a silicone coating is as follows:

Part A	*Part B*
Silicone polymer	Cross-linker
Solvent (optional)	Catalyst
	Solvent (optional)
	Dyes (optional)

Property Profile

Easily applied, like two-component urethanes and epoxies
Cure slowly, heat usually necessary
Very difficult to repair; must be mechanically removed
Short pot life
Flammable solvents but higher flash point than acrylics
Very good chemical resistance
Very good solvent resistance
Very good short-term humidity resistance
Good long-term humidity resistance
Good abrasion resistance
High-temperature resistance best of the four
Primer recommended
Highest cost

Use Considerations. These silicones should not be confused with the soft room-temperature vulcanizing (RTV) silicone rubbers of Chapter 3, which are used in potting. These are more resinous and harder in nature. Their forte is high temperature resistance and a wide service range, from approximately 60 to over 200°C, which is usually beyond the capabilities of the components and the board they are protecting. Removal is a problem, as they resist most removers and their temperature resistance makes soldering through difficult or impossible.

D. Parylene

1. Composition

Parylene is not a formulated material. There are three versions available from Carbide, all based on polyparaxylylene, known as Parylene N. Parylene C has one Cl added to the ring. Parylene D has two.

2. Property Profile

An absolutely uniform coating over all surfaces, edges, points, and under components, but does not bridge holes. It is a line-of-sight coating process.

>   Absolutely uniform coating over all surfaces regardless of configuration
>   Does not bridge holes or component-to-board interfaces
>   Very abrasion resistant
>   Instant cure
>   Substrate never sees above-ambient temperature
>   Excellent chemical and solvent resistance, almost unremovable
>   Good heat resistance in air
>   Very high capital cost
>   Thickness from 0.25 to 3.0 mil with single coat
>   Excellent dielectric properties
>   Extremely difficult to remove for repairs
>   Very low thermal stress on components or leads

3. Use Considerations*

As described in Chapter 4, Parylene is a unique material requiring a unique process. Since the vapors are pervasive, areas must be masked if they must be coating-free. The capital cost—in excess of

---

*For more information, see Ref. 5.

$25,000—and the high cost per applied mil make Parylene usually the last consideration despite its superior properties. *Circuits Manufacturing* [6] described it as the Rolls-Royce of conformal coatings, and this analogy is apt. Repair is difficult, but protection is superior, so where the ultimate in environmental protection is required on a board or component, it is Parylene.

E. Materials of the Future*

Recently, ultraviolet (UV) curable resins have become available. The advantage they offer is almost instant cure, effected by the use of high-intensity UV light of a prescribed wavelength. However, high shrinkage has been a problem, potentially leading to component, and, in the case of surface mount, bond line stress. Also, the shadowed areas, those not in the direct line of the radiation, do not cure, leading some conformal coating suppliers to build a secondary mechanism into their systems. However, these secondary mechanisms tend to negate some of the "instant" advantages of UV coating, resulting in a two-component pot life situation or a post-heat cure.

Most UV systems are 100% solids, eliminating the solvent hazard and the energy-consuming step of solvent removal.

Hybrids of acrylics and urethanes are also available now to improve the weak points of each of the original materials, the urethane contributing flexibility and improved abrasion and solvent resistance, and the acrylic contributing moisture resistance and the UV cure mechanism. Even acrylic/epoxy alloys are available. Unfortunately, these, like the pure acrylics, are expensive due (in no small part) to the cost of the photoinitiators. So far a pure epoxy (room temperature), UV cured, does not exist as trace impurities inherent in the epoxies inhibit the initiators.

When considering a UV cured conformal coating, the higher material cost must be balanced against the very fast through-put rate and low energy consumption of the UV source.

---

*For more information, see Ref. 7.

F. Safe Handling

Conformal coatings should be used only in well-ventilated areas and away from sparks and flames, as almost all of them contain solvents. Avoid inhalation of vapors. Store in a cool place and ground both containers when pouring material. This is the first time we have encountered solvents in encapsulation materials, and great caution is necessary.

G. Remember

The encapsulant selected to protect an electronic device or component is seldom the ideal or even the best one available. It is a compromise, taking into account all the "cans" and "cannots," "wills" and "will nots," people available and their capabilities and facilities, money available, and performance of the device. Be open with your material supplier—use his experience; it is free.

REFERENCES

1. Impregnation study of porous elements of high voltage components, IEEE *Trans. Electr. Insul.* Sept. 1985.
2. H. S. Katz and J. V. Milewski, eds., *Handbook of Fillers and Reinforcements for Plastics*, Van Nostrand Reinhold, New York, 1978.
3. Public Health Service Publication No. 1040 5/63, Superintendent of Documents, U.S. Printing Office, Washington, D.C.
4. How to Use Conformal Coatings, in *Printed Circuits Handbook*, C. F. Coombs, Jr., ed., McGraw-Hill, New York, 1967.
5. Union Carbide bulletin, "Parylene—Environmentally Compatible Conformal Coating."
6. The overcoat, *Circuits Manufacturing*, Sept. 1983.
7. Coatings Cover High-volume PCBs — *Electronic Packaging & Production* — Feb. 1986.

APPENDIX: SOURCES FOR MATERIALS

Liquid Epoxy Compounds

Acme Division, Allied Products, New Haven, Connecticut
Amicon Division, W.R. Grace, Lexington, Massachusetts
Armstrong Products, Warsaw, Indiana
Bacon Industries, Watertown, Massachusetts
Castall, Inc., East Weymouth, Massachusetts
Conap, Inc., Olean, New York
J.C. Dolph, Monmouth Junction, New Jersey
Emerson Cummings, Canton, Massachusetts
Epic Resins, Waukesha, Wisconsin
Epoxy Technology, Billerica, Massachusetts
Formulated Resins, Greenville, Rhode Island
Furane Products, Los Angeles, California
H.V. Hardman Co., Belleville, New Jersey
Hysol Division, Dexter Corp., Industry, California
Insulcast Division, Permagile Salmon, Ltd., Plainview, New
    York
Isochem Resins, Lincoln, Rhode Island
Mereco Products, West Warwick, Rhode Island
Ren Plastics, East Lansing, Michigan
Rexnord, Commerce City, Colorado
Ripley Resins, Addison, Illinois
Smooth-on-Manufacturing, Gillette, New Jersey
Thermoset Plastics, Indianapolis, Indiana
3M Co., Minneapolis, Minnesota

RTV Silicones

Castall, Inc., East Weymouth, Massachusetts
Dow-Corning Corp., Midland, Michigan
General Electric Co., Waterford, New York

Insulcast Division, Permagile Salmon, Ltd., Plainview, New York
McGhan Nusil Corp., Carpenteria, California
S.W.S. Silicone Corp., Adrian, Michigan
Transene Co., Rawley, Massachusetts

Liquid Polyurethane Compounds

Castall, Inc., East Weymouth, Massachusetts
Conap, Inc., Olean, New York
Emerson & Cummings, Canton, Massachusetts
Formulated Resins, Inc., Greenville, Rhode Island
Furane Products, Los Angeles, California
H.V. Hardman, Belleville, New Jersey
Hexcel Corp., Chatsworth, California
Hysol Division, Dexter Corp., Industry, California
Insulcast Division, Permagile Salmon, Ltd., Plainview, New York
Isochem Resins, Lincoln, Rhode Island
Mereco Products, West Warwick, Rhode Island
Ren Plastics, East Lansing, Michigan
Thermoset Plastics, Indianapolis, Indiana
3M Co., Minneapolis, Minnesota

Liquid Unsaturated Polyester Compounds

Normally, base polyester resin is purchased and the limited formulating that is necessary is done in the user's facility.

Freeman Chemical, Chicago, Illinois

Compression Molding Powders

F.S.C. Co., Riverside, California
Hysol Division, Dexter Corp., Industry, California

Pennwalt Corp., Philadelphia, Pennsylvania
Plaskon Electronic Materials, Toledo, Ohio
Plastics Engineering Co., Sheboygan, Wisconsin
Reichhold Chemicals, North White Plains, New York
Thermoset Resins, Indianapolis, Indiana

Transfer and Injection Molding Powders

F.S.C. Co., Riverside, California
Hysol Division, Dexter Corp., Industry, California
Plaskon Electronic Materials, Toledo, Ohio
Thermoset Plastics, Indianapolis, Indiana
Tool Chemical, Ferndale, Michigan

Bulk Molding Compounds — B. M. C.

B.M.C., Inc., St. Charles, Illinois
Enplax Corp., Belleville, New Jersey
The General Industries Co., Elyria, Ohio
Glastic Co., Cleveland, Ohio
Haysite Reinforced Plastics, Erie, Pennsylvania
Plaskon Electronic Materials, Toledo, Ohio
Premix, Inc., North Kingsville, Ohio
Tool Chemical, Ferndale, Michigan

Conformal Coatings*

(ER, AR, UR), Conap Corp., Olean, New York
(SR), Dow Corning Corp., Midland, Michigan
(ER, AR, UR), Furane Products, Los Angeles, California
(SR), General Electric Co., Waterford, New York
(ER, AR, UR, SR), Humiseal Division, Columbia Chase Corp.,
    Woodside, New York

---

*ER, epoxy; AR, acrylic; UR, urethane; SR, silicone; PR, Parylene.

(ER, UR), Hysol Division, Dexter Corp., Industry, California
(ER), Insulcast Division, Permagile Salmon, Ltd., Plainview,
New York
(SR, ER), Transene Co., Rawley, Massachusetts
(PR), Union Carbide, Danbury, Connecticut

6

Preparing to Encapsulate

Tom Baker
Principal Engineer, Equipment Division Laboratories
Raytheon Corp., Sudbury, Massachusetts

I. INTRODUCTION

Several years ago, the term "encapsulation" had a distinct difference in meaning from embedment, potting, encasement, casting, and impregnation. While "impregnation" still means the application of a resin to fill porosity in a device (transformer windings, metal castings, etc.), the other terms have become almost synonymous. Today, these terms mean pretty much the same thing—the enclosure of a device, assembly, subassembly, and so on, with a resin system. Little distinction in nomenclature is now made whether the encapsulation is accomplished with the unit in a removable mold or in a mold that becomes an integral part of the final encapsulated unit or assembly. Therefore, the term "encapsulation" will be used here in the context of completely covering an electronic device or assembly using a pourable, curable, liquid, electrically insulating, dielectric resin system.

II. PREPARING TO ENCAPSULATE

Now that you have selected an encapsulation material for your design, you are ready to encapsulate. Well, almost. Assuming that the encapsulant to be used has been correctly selected based on the design requirements of the unit or assembly to be encapsulated, there are several factors that still must be considered beforehand. The actual step of encapsulation can be a very simple one or it can be one requiring a meticulous, systematic processing sequence that must be followed religiously. Perhaps the most demanding encapsulation is that which requires a void-free product such as is needed when encapsulating high-voltage power supplies. In addition to being void-free initially, encapsulants must also have excellent adhesion to all components within such power supplies. This is particularly true for high-voltage power supplies that are intended for airborne applications. Unbonded encapsulant will act as a void at altitude and because of the laws of dielectric breakdown could be the source for corona and eventual failure of the unit. The following step-by-step procedures are typical of those required for producing void-free encapsulation as obtained by vacuum processing. If your design is less complex and less demanding, it will probably not be necessary to follow each step. The decision as to which steps to eliminate should become clear as you follow through these procedures. It will depend on the ultimate design objectives that you are trying to achieve.

A. Compatibility

As a general rule, compatibility of the encapsulant (and primer, if used) with the materials of construction of the assembly must be established prior to encapsulation. The more critical the intended application of the assembly (such as high voltage at altitude), the more necessary it is that the encapsulant is compatible and that it adheres to all surfaces.

If fluorinated plastics such as Du Pont's Teflon are used for wire insulation or insulated terminals, these must be specially treated, since nothing will adhere to them. Fortunately, there are commercially available etchants that will render the surface of fluorinated plastics bondable. AMS2491 is one specification that gives a general procedure on how to accomplish this. What this means, however, is that fluorinated plastics must be surface prepared prior to use within the assembly. For example, wire insulation must be treated in the special etchant before it is cut to the size required for wiring in the assembly. Since the etchant could possible wick into the ends of an insulated wire, it is necessary to keep the ends out of the etchant, and subsequently the unetched portions should be cut off and discarded. It is also possible to buy preetched fluorinated plastic insulation wire.

In either case, however, the treated surface is adversely affected by ultraviolet radiation. Adhesion to the specially treated surface will degrade over a period of time if left unprotected from ultraviolet radiation. Therefore, it is good practice to keep treated insulation wrapped or stored in some noncontaminating material such as kraft paper.

Other wire insulations, or for that matter, all materials used within the assembly should be spot tested for compatibility and adhesion with the encapsulant. One special word of caution is needed if the encapsulant is an addition-cured silicone. These materials are particularly sensitive to certain other materials, such as vinyls, condensation-cured silicones, certain amine-cured epoxies, and others. If not compatible, the addition-cured silicone in contact with these materials will not cure; they are inhibited from curing, and will stay wet and sticky. Obviously, compatibility tests will be required. Special attention must also be given to silicone insulation on wires. Remember that nothing sticks to a silicone except a silicone. Epoxies and urethanes will not stick to silicone-insulated wire. In fact, many silicones also do not readily adhere to silicone-insulated wire. For high-voltage applications it is necessary to test to ensure compatibility. In the case of silicone-insulated wire to be used in a silicone encapsulant, it will be necessary to run an extended test. Some silicones that appear compatible and do adhere to certain silicone-insulated wires initially are affected over a period of time. There is a possibility of residual peroxide catalyst in the wire migrating to, and actually embrittling, the encapsulant adjacent to the wire. I do not discourage the use of silicone-insulated wire but only point out that if it is to be used, it must be thoroughly tested for long-range compatibility with the encapsulant.

B. Cleaning

As a general rule, all parts should be cleaned prior to the encapsulating. In particular, solder flux, metallic debris such as chips, and fingerprints should be removed. As with other operations, the more critical the end application of the assembly, the more necessary it is to ensure the adequacy of the cleaning process. Improper surface preparation and cleaning are the major causes of subsequent failures of encapsulated electronic assemblies. If possible, cleaning should be performed immediately before the encapsulation step to ensure that the assembly is clean at that point. Solvent cleaning is preferred using a solvent or combination of solvents such as trichloroethane, isopropyl alcohol, or one of the fluorinated azeotropic mixtures. It should also be established whether vapor degreasing will be required rather than cold solvent cleaning. It should also be predetermined that the solvents used are compatible with all the materials of construction within the assembly to be encapsulated. Certain plastics might be attacked by the solvents, or elastomers may swell to the point of irreversible damage. It will also be necessary to establish if a baking or vacuum baking cycle is needed after cleaning prior to encapsulation. A typical cleaning process employed in the electronics industry might consist of vapor degreasing, deionized or distilled water rinsing, alcohol rinsing, and baking operations. Once parts are cleaned they should be handled with care. Lint-free gloves are recommended for handling. Rubber or plastic gloves are less preferred since some—

not all—can leave residues on the cleaned parts. If questions arise on this, the nature and compatibility of the gloves should be checked.

These cleaning steps should result in a clean assembly or unit to be encapsulated. However, if good adhesion of the encapsulant is required, solvent alone will not usually achieve it. This becomes especially critical if the encapsulant is to be a silicone. It is absolutely necessary that where high adhesion is required, it will be necessary *prior* to the solvent cleaning operations to roughen (i.e., scuff sand or sand blast) all glossy surfaces. Molded plastic parts usually have a high glossy surface and often, some residual mold release is still retained in these surfaces. It is necessary to solvent clean and scuff sand these surfaces to prepare them properly for encapsulation.

If the assembly is to be encapsulated and contained within an aluminum box, the aluminum should be presanded and cleaned prior to the assembling of components within it. This is necessary to break the mill finish surface typically found on wrought aluminum boxes and containers. If the aluminum container has already been chromated, a decision on surface preparation must be made. There has been a controversial issue for many years on the subject of bonding and adhering to chromated aluminum surfaces. Assuming that the chromate film has been properly applied and is not too thick or powdery, the question of adhesion is as follows. When bonding to a fresh chromate (i.e., up to 4 days old), adhesion on the order of freshly sanded or chemically etched bare aluminum can be obtained. As the chromate film ages, it reacts with moisture and self-seals itself. The older chromate becomes quite smooth, and there is little porosity to provide a high surface area for bonding. When bonding to older chromate, adhesion values can be as low as 50% that of a fresh chromate. Therefore, the question arises as to whether or not the old chromate should be removed by scuff sanding prior to bonding (i.e., encapsulating in this case). High initial strengths can be obtained if this is done. This sounds as if the decision should be easy; however, if the encapsulated unit is to be exposed to a high humidity environment, there is another consideration. The durability of a bond to a chromated aluminum can be orders of magnitude better than to a sanded or etched aluminum surface. Although the initial adhesive strength to old chromated aluminum is lower, it will retain most of this strength under adverse humidity conditions. The higher initial strength to the sanded or etched aluminum will decay very fast under adverse high temperature/high humidity conditions. The severity depends on the aluminum alloy since some are more prone to humidity attack than others.

Remember that the discussion above applies if the encapsulated unit is to be exposed to severe environmental conditions. For most military electronics, this will usually be the case. Because of the wide temperature and humidity conditions that military electronics might possibly see, it is normally required that highly durable bonded surfaces be obtained. If there are questions regarding how good your surface preparation-cleaning-encapsulating processes are, the only answer is to run tests to establish the needed data.

C. Priming

The priming operation is best described as a nuisance. It is time consuming, usually messy, involves the handling of high solvent materials, and is an operation that is difficult to inspect for results. Unfortunately, when a primer is required, the priming operation ranks just behind surface preparation and cleaning as the most critical. Fortunately, epoxies and most urethanes do not require the use of a primer since their adhesion is usually adequate. Silicone encapsulants, however, do require priming to obtain adhesion. Recent technology advancements have resulted in the new self-priming silicone encapsulants; however, this technology has not yet developed to include all silicones.

When priming is needed it should be done as soon after cleaning as possible. Simple parts may be sprayed or dipped, but these techniques are inadequate for densely populated assemblies such as a high-voltage power supply assembled within its container. In this case, it usually means that the primer has to be poured into the unit, sloshed around to coat all surfaces, and then poured out.

As stated above, it is messy, but there is no better way of ensuring that all surfaces within the unit to be encapsulated have been coated with primer. It is necessary for adhesion purposes to keep the actual thickness of the primer as thin as possible since thick primer will usually fail cohesively within itself when strained. Typical primers for silicone encapsulants have a solids content below 10%. Even at this low level, the application of primer by pouring in and then out can leave pockets of thick primer coating. One technique developed over the years is to add additional solvent to the primer to reduce solids content even further, to allow for more uniform application. I cannot emphasize enough how important it is to get good uniform primer coverage, particularly for critical high-voltage applications. The priming operation will require special attention to each and every application to ensure adequacy.

I mentioned previously that priming should occur immediately after cleaning, but I did not say whether to prime after cleaning and baking or after cleaning but before baking. The exact sequence will have to be established for each specific application. Normally, it is possible to clean, prime, allow the primer to cure, and then bake (or vacuum bake) without any deleterious effect upon the primer. This is the preferred way of doing it; however, it may not be possible if the primer is affected by the baking cycle. The only way to know for sure is to conduct tests to establish the optimum sequence.

At this point, the assembly to be encapsulated is cleaned, primed (when required), baked, and should be ready for the encapsulation process.

D. Testing Before Encapsulation

Perhaps one of the best innovations in high-voltage encapsulation over the last decade or so is the ability to test a high-voltage assembly just prior to final encapsulation. This is now possible using an inert dielectric fluid such as a fluorochemical. It is therefore possible to submerge an assembly in this fluorochemical and to run the unit electrically to determine its adequacy. It is amazing how many problems this technique will uncover and obviously it is better to find these problems at this stage rather than after encapsulating. As mentioned earlier, these fluorochemicals are almost completely inert and leave no residue with the assembly. These materials are relatively expensive; however, they pay for themselves many times over. One word of concern, however; being liquid, fluorochemicals have better heat transfer capabilities than do solid encapsulants, because of convection. This fact should be considered if the thermal response of an operating functional high-voltage assembly is important.

III. ENCAPSULATING

A. The Encapsulation Process

As mentioned previously, the actual encapsulation step can be very simple or it can be quite complex. The degree of difficulty is directly related to the ultimate requirement for the finished encapsulated unit. Assuming that everything is now all ready for the final encapsulation step (for high-voltage vacuum encapsulation it really isn't yet), let's proceed with the various possibilities that are available.

B. Type of Encapsulant

If the encapsulant is one that requires heat to effect its cure, a long working time (pot life) at room temperature can be expected. If the encapsulant is one that cures at room temperature, the major concern becomes one of time. Remember that room-temperature curing materials, usually two-part systems consisting of resin and curing agent (or a catalyst), will start the reaction process immediately upon mixing. The working time of the encapsulant must be within the time frame needed to complete the encapsulation step.

C. Simple Encapsulation

The simplest encapsulation is to mix the encapsulant and pour it into the assembly or mold. If it is a one-part heat-cured encapsulant, it is just as simple—pour it into the unit and place it into an oven to cure. This sounds simple and it is. The majority of the effort has already been accomplished in the surface preparation and cleaning steps. The encapsulation step itself requires little effort if you are not concerned with voids or bubbles in the encapsulant. One word of caution, however. Even with this simple step, it is necessary to give some attention to the encapsulant prior to using it. Many encapsulants contain fillers that provide special features, such as improved thermal conductivity, reduced coefficient of expansion, reduced shrinkage, and so on. For two-part systems, the container with the fillers—usually the resin but sometimes also the curing agent—should be prestirred prior to weighing out to ensure uniformity. Fillers settle, and because of this the material should be stirred in its container each time prior to usage.

D. Material Evacuation

If some voids or bubbles are tolerable within the cured product, but it is desired to remove most of them, the mixed encapsulant should be degassed prior to pouring. The encapsulant should be in a container that has at least five times more volume than is occupied by the encapsulant itself. The degassing step is usually done in a belljar. The belljar should be adequately protected with a safety shield to protect workers. As the pressure decreases (vacuum increases) the air in the encapsulant will cause the encapsulant to rise. If there is not enough free space in the container to accommodate this, the material will overflow the container and make a mess. One way of avoiding this is to "play the vacuum." This means that as the encapsulant rises to the top, you have to adjust the vacuum, usually with a valve, to hold the encapsulant from rising over the top. Letting a little air back into the belljar for a second or two will force the bubbles down; continued vacuum will then cause the encapsulant to rise to the top. If you have never done this before, it can be a little tricky. With a little experience, it almost becomes second nature to control the encapsulant where you want it. Once the large bubble collapses and the material subsides into the container, the major degassing step has taken place. Continued vacuum will help remove some additional air; however, if continued too long it can be detrimental. First, valuable working time of the encapsulant is being used up for room-temperature curing systems. Second, extensive vacuum treatment can actually strip off low-molecular-weight fractions from the material. This could alter the cure of the material and possibly its end-product physical properties.

One word on the vacuum system itself is in order. The vacuum system should be of sufficient capacity to reach about 5 torr (1 torr = 1 mmHg) within a few minutes. If it takes longer than this, again, valuable working time will be used. Another word of caution is that a thoroughly adequate vacuum pump can become very inefficient if the actual plumbing into the belljar or vacuum chamber is too restrictive. The final pipe or tubing into the vacuum apparatus should not be less than ½ in. inside diameter if possible.

For difficult to degas materials, it may take many minutes to break the main air bubbles. If this is too long in time, a larger container may be helpful. Warming the material will reduce its viscosity and allow for faster degassing. The addition of heat to a room-temperature curing material, particularly one that can exotherm such as an epoxy, will normally reduce its working time. Therefore, be sure to check this beforehand if it is to be part of the encapsulation procedure.

E. Material Degas: Assembly Degas

Degassing the encapsulant as previously described, pouring it into the assembly to be encapsulated, and then degassing the encapsulant in the assembly usually will help in further reducing bubbles or air entrapment. Unfortunately, sometimes the opposite is found, wherein degassing the encapsulant in the assembly will generate bubbles from air that is entrapped in crevices or from other materials, such as PC-board assemblies. Prevacuum bake of the assembly is helpful in reducing some sources of bubbles.

Although this technique is probably the one used most within the electronics industry for encapsulation, it is not the ultimate way. The best known method to minimize or to completely eliminate voids or bubbles is that of encapsulating while *both* the encapsulant and the assembly are under vacuum prior to pouring. This is the ultimate technique for the encapsulation of high-voltage equipment.

F. Vacuum Encapsulation for High-Voltage Equipment

If the assembly to be encapsulated is already under vacuum with the encapsulant (prior to pouring), most sources that could generate air or gas should be eliminated or at least greatly reduced. The pouring of the degassed material into the degassed assembly should lead to minimal bubbles. Unfortunately, this is not the case unless other special precautionary measures are taken. Remember that it is assumed that the assembly has previously been prevacuum baked and cooled to room temperature (for use with room-temperature curing encapsulants). The vacuum system should be stabilized in the range 2 to 5 torr before introducing the encapsulant. A harder vacuum may be drawn and then returned to, and stabilized at, 2 to 5 torr as part of the procedure. The encapsulant should be exceptionally quiet at this point (i.e., no hiccups of bubbles being generated in the encapsulant). However, when the encapsulant is then introduced into the assembly, the stream of encapsulant may now start to bubble. The reason for this is the vast increase in surface area that the encapsulant now offers in the form of a stream compared with when it was poured from a container. Remember, however, that these bubbles are at 2 to 5 torr (or whatever other vacuum is being used). These bubbles should collapse and disappear upon subsequent release of air back into the vacuum chamber. This will happen only if certain precautions are built into the encapsulating process. First, there should be a reservoir for the encapsulant built into the mold of the unit being encapsulated. When the assembly being encapsulated is filled, and the reservoir is filled enough, the release of air into the chamber will act on the reservoir and force the encapsulant down into all unfilled interstices of the assembly. This is the same principle as a hydraulic pump. This is effective only as long as there are no other locations for air to get back into the assembly. Thus the mold or assembly must have been presealed with some removable sealant during the preparation of the assembly for encapsulation; this means far back in the process even before going into the vacuum bake.

Following these procedures should lead to a void-free high-quality encapsulation. It should be noted that some manufacturers will actually go one step beyond the vacuum encapsulation. A final pressurization step is added in which immediately after vacuum encapsulation, the assembly is pressurized to help drive the encapsulation further into any remote interstices. This procedure is generally performed when encapsulating transformer coils, where it may be difficult to replace all entrapped air by vacuum alone. Once again, testing is the only way to verify exactly what processes will be needed for a specific application.

IV. SOME OTHER CONSIDERATIONS

A. Room-Temperature Curing Materials

All other things being equal, room-temperature curing material is generally preferred since there is essentially little stress induced into the encapsulated assembly at room temperature. Thus thermal excursions to high or low temperatures are minimized. A material that requires high-temperature curing will mean that at room temperature, stresses from shrinkage and thermal expansions exist in the assembly. Further exposure to low temperature will further aggravate and increase these stresses. These stresses are a function of $E \cdot \alpha \cdot \Delta T$ (modulus $\times$ coefficient of thermal expansion $\times$ temperature differential). Curing at room temperature helps to minimize the overall ΔT if the encapsulated unit is to be exposed to wide temperature ranges. It is also a good rule that materials should be cured at a temperature slightly higher than it will experience in actually operation (and sometimes storage) conditions. Therefore, the best guideline to follow if possible is to allow the room-temperature curing material to cure for at least 16 hr (usually overnight) and then postcure it at some temperature slightly higher than its eventual environment temperature. Rather than postcure directly at the high temperature, it is also advisable to postcure in steps by allowing the encapsulant to experience intermediate temperatures for some reasonable time.

B. Elevated-Temperature Curing Encapsulants

When using an elevated-temperature curing encapsulant, it should be realized that absorbed gases, including air and carbon dioxide, are less soluble the higher the temperature. This will have an impact on voids and bubbles in the cured encapsulant if not fully addressed. If, for example, an elevated-temperature curing encapsulant was poured cold (i.e., at room temperature) into a cold mold and then placed into an oven to cure, bubbles would be liberated from the encapsulant. As the temperature of the encapsulant is increased, absorbed gases—all of which do not necessarily come out during vacuum degassing—will be liberated as their solubility decreases. It is possible and often probable that some of these bubbles will be entrapped as the encapsulant cures. This is, of course, intolerable for high-voltage encapsulation. To overcome this, the unit to be encapsulated and the encapsulant itself during the encapsulation process shoud be at a temperature slightly higher than the cure temperature. The slightly higher temperature will help compensate for thermal loss during the time it takes to encapsulate. Although the procedure sounds somewhat excessive, it is in fact needed if void-free encapsulation is the ultimate goal.

C. Corona Ground Shield

Some applications require that a high-voltage assembly be totally electrically isolated from its surroundings. Being an insulator, the encapsulant will obviously not achieve this. It is therefore necessary to coat the exposed surfaces of the encapsulant with a low-conductivity or resistive coating. This coating must be compatible and adhere to the encapsulant. It will also be necessary to have this coating bridge to, and adhere to, the metallic surfaces adjacent to it. When this is accomplished, the high-voltage assembly is now essentially within a Faraday cage, such that it operates in an all-grounded environment.

D. Guidelines for High-Voltage Encapsulation

It is beyond the scope of this book to get into the myriad of do's and don'ts required for successful encapsulation of high-voltage assemblies. It is somewhat surprising that a well-organized treatise on this subject does not exist.

There are many ground rules for successful high-voltage encapsulation. I will mention only what I feel is the most important here and leave it to the reader to research the subject to the extent required. In my mind, the most important rule that governs the encapsulation of a high-voltage assembly is this—the assembly must be initially designed for encapsulation. The fact that an assembly will function properly if filled with another dielectric material such as oil or gas is no guarantee that it will function similarly if encapsulated. Therefore, it is imperative that the designer know the ground rules for high-voltage encapsulation and apply them to the design during the initial design and layout stages. This is an absolute necessity.

7
Troubleshooting

I. INTRODUCTION

If all encapsulation problems were listed and their ultimate causes determined, experience shows that 90% or more would be traced to a cause *other than* encapsulant performance, although at the time of the problem, more than 90% would be blamed on the encapsulant material. Wrong utilization usually proves to be the real culprit.

At this point you have selected the material to be used, applied it by the appropriate method, and now:

1. Your encapsulated device is not cured.
2. The encapsulant is cracked.
3. The encapsulant is shrinking away.

4. There are soft and hard spots in the potted unit.
5. The conformal coating has soft areas.
6. The molded resistor is not completely covered.
7. The molded coil is "out of round."

All of the above and more can, *and do*, occur frequently in encapsulation processes. So we will look at some of the more common problems and their probable causes. But first, some simple advice. It may sound trite, even obvious, but you will ignore it at your peril.

If encapsulated units are not hardening, filling, or coating correctly, the first thing to do is: *Stop filling, casting, molding, or coating more units until the cause has been determined and corrected.*

II. CLASS I MATERIALS

A. No Cure

When a unit has been poured and has not hardened after three to four times the stated pot life, the most likely cause is too little, or no, curing agent used.

1. *Hand-mixed materials.* Check the weights and ratio of parts A and B. If this is inconclusive, supervisory personnel should mix a small test batch and accelerate the cure with heat. If it cures, the cause is obvious—the hardener was not added.
2. *Dispenser in use.* First check the hardener reservoir. Is it empty? If not, is it being pumped. The best check to run now would be to bypass the mixer and run the dispenser, collecting the base and hardener in separate, preweighed containers to check the ratio. If not correct, readjust after determining why it drifted off. Is a pump malfunctioning? A valve leaking or not opening? A line plugged or restricted? A standard daily startup procedure (or after

any shutdown) should be to shoot into a test container and allow it to harden, even accelerating it with heat if necessary to reduce shutdown time. If within established limits, start filling units.

B. Cures Too Fast

Consequences of too-fast cure are higher-than-normal exotherm temperatures, high apparent shrinkage, and very poor flow across the unit and around components. In extreme cases, bubbling and lavalike appearance of the encapsulant results. This is an extreme case of runaway exotherm. High-temperature rise is not a silicone symptom. Check the following.

1. *Ratio of base to reactor.* In room-temperature systems, very high reactor amounts cause faster cures and higher temperatures.
2. *Temperature of starting materials.* As little as a 15°F increase in material temperature will speed up the gel time (shorten hardening time) by as much as one-third in small units and one-half or more in larger units. Each May/ June a "rite of spring" is celebrated. Warm nights follow hot days and material temperatures rise. The next day encapsulants are suddenly "hardening much faster," particularly medium to slow room-temperature systems. Check the plant temperatures of the preceding 2 to 3 days, particularly a warm weekend, then check material temperature.
3. *Oven temperature.* If a heat-cure system is in use, check oven temperatures. These frequently drift off or are changed by someone unauthorized to do so.

If none of these are the cause, call your material supplier.

C. Cures Too Slowly

This presupposes that it eventually does cure.

1. Check as in item 1 in Section II.B. Much too low a hardener ratio will greatly slow the hardening process.
2. Check as in item 2 in Section II.B. A drop of as little as 15°F in temperature will greatly extend the hardening time, particularly in small units. The smaller the mass of resin, the greater the slowdown. Each October/November, nights cool off and so do materials. So hardening times "suddenly" increase. One cold night in, say, New Jersey, Pennsylvania, or Ohio, giving plant temperatures in the 50s, will cool materials to that temperature. With the heat "on" the next day, the ambient will quickly rise to ±70°F, but not the material. It will stay low, resulting in slow hardening. Liquid encapsulants "suddenly" not curing fast enough in October/November are the "rite of autumn."
3. With room-temperature systems, check their location as they cure. A drafty door or window unnoticed in warm weather can chill the units into two or even three times normal turnover times in the cooler months.
4. Check the mixing efficiency of both hand and dispenser systems. Poor mixing will cause slower and even incomplete hardening. A symptom of this is variation in hardness across the surface.

If none of the above, call the material supplier.

D. Cured Materials Have Softer Spots (or Harder Spots)

There are two things that can cause this phenomenon.

1. *Incomplete mixing.* If the reactor and base are mixed for too short a time, there will be areas that have an excess of base and areas with excess reactor. These areas will cure slowly and remain soft and incompletely cured, or, in extreme cases, not cured at all. The solution is to mix longer. It is sometimes possible to have a contrasting

color dispersed in the reactor, so incomplete mixing will show up as color swirls, as in multiflavor ice cream. If a dispenser is in use, check the mixing tube. Is it properly packed? Is it large enough? If newly set up, perhaps a longer tube is necessary.

2. *Improper mixing technique.* It is possible to get soft spots even if the base and reactor have been mixed to a uniform color. This usually shows up as one or two units out of a large batch having a soft area, usually a line or spot in the middle of the top surface. This is usually a hand and mechanically mixed phenomenon, not a dispenser mix. Check to see that the sides and bottom of the mix container were scraped to incorporate all the material. This unmixed side and bottom material will have very little reactor in it, which will show up as a soft or uncured area. It usually shows up in the last one or two units filled, as this is the last material to come out of the container, particularly if the sides and bottom are scraped to remove this residual material, and it usually ends up toward the middle of the top surface. If a mechanical mixer is used, some provision for scraping the sides and bottom (during mixing) must be made. If the mixes are done by hand, technique must be improved.

E. Inconsistent Cure

There are two most probable causes.

1. Units start out curing rapidly, then, subsequently, days or weeks later, units cure more and more slowly, usually accompanied by increasing viscosity of the base component and mixed material. Check to see if the encapsulant has undergone filler settling in the original container before any is withdrawn for mixing. When filler settles, the top of the container becomes rich in the resin (silicone, urethane, epoxy) component. As material is removed, more resin than filler is used. Toward the bottom, more filler

than resin is used, giving higher viscosity and slower cure. The solution is to premix material in its original container before withdrawing any. Where a dispenser is used, a mechanical mixer can be a part of the material reservoir.

With epoxies, filler settling can also cause "off-ratio" conditions with the reactor even though correct measurements were taken. This is particularly true when a low-ratio, unfilled curing agent is used, DETA for instance. The reactor ratio for the base is preset. When filler settling causes the material at the top of the container to be enriched in epoxy, the preset amount of hardener will be too little to cure it correctly. Slow cure and property development will result. The reverse will be true when approaching the bottom of the container and viscosity rises and gel accelerates. Very poor electrical properties are also a symptom here. Many other problems can emanate from this one; for example, lower filler concentration near the top of the container can cause higher shrinkage, lower thermal conductivity, and degraded thermal shock resistance. Higher filler level causes air retention and poor flow.

2. The encapsulant cures to a different hardness and at a different speed than that of previous samples. This is usually found with meter-mix-dispenser equipment. Check for ratio setting and air locks in the cylinders. This is often a startup problem with new equipment.

F. High Shrinkage (Encapsulant Pulls Away from the Case or Mold During Cure)

This phenomenon can be caused by incorrect material selection, particularly in larger units more than 1 in. in depth and 3 in. or more in diameter. Check two things:

1. The encapsulant selected may shrink too much for the size of the unit during its cure. This might be the case

with an unfilled material and a large unit or with a base material with a low filler level and a very high ratio of reactor, which, in effect, lowers the net filler level. Changing to an encapsulant with lower inherent shrinkage will correct this.

2. With epoxies and urethanes, the system may be too fast for the size of the casting and cause excessive temperature rise and "induce" high shrinkage. If the system selected is not suited to the size of the unit, the heat produced during cure, either room or elevated temperature cure, is generated much faster than it can be dissipated, causing thermal expansion of the encapsulant. As the cure accelerates, more heat is evolved and rate increases, the temperature rises, and thermal expansion increases. This scenario continues until gelling or hardening occurs, accompanied by an additional temperature rise. This hardening has occurred with the encapsulant in a greatly thermally expanded condition. As the unit cools to room temperature, the encapsulant now has to accommodate thermal shrinkage, as it, too, returns to room temperature. The cooling process generates stresses in the encapsulant, which are relieved in one of three ways:

   a. The encapsulant pulls away from the side of the can or mold. If a mold, the casting may be out of shape and cosmetically unacceptable. If a can, a point of entry for contaminants is provided between the can and the cured material. This shrinkage can also damage brittle or pressure-sensitive components, break wires, and open solder connections.

   b. If the adhesion of the encapsulant is sufficient, the sides of the can will be pulled in, sometimes to the point of fracture. This pull-in can be severe enough to be cosmetically unacceptable.

   c. The encapsulant will crack or split, causing a fissure or crack. When this occurs, something inside has surely been damaged, if not destroyed.

The solution to thermally induced high shrinkage is simple. With room-temperature cure systems, switch to a slower one suitable to the mass of material being cast. For high-temperature cures, try a lower oven temperature to achieve gelation, then elevate if *necessary* for further encapsulant property development. A slower curing agent could also be used here. More than two-thirds of the shrinkage attributable to cure takes place while the material is still liquid (1). Allowing as much of this shrinkage as possible to take place before the hot spot—the middle of the casting—gels results in much less built-in stress, which increases as the casting cools.

   Special note.   All encapsulants shrink on cure. The shrinkage referred to here is that which is above and beyond the inherent shrinkage in any system and which is induced by improper material selection and use.

G. Epoxy Encapsulant Hard but Brittle (Shatters Like Glass When Impacted)

Check the curing agent. Determine if it is an aromatic amine. These curing agents require some heat to complete the cure. If insufficient heat or time is allowed, aromatic amines go to a first or B stage cure. Although hard at room temperature, B stage epoxies are very brittle, with a low softening point, below 125°F. The remedy is to raise the cure temperature and/or extend the time. From 12 hr at 140°F to 2 hr at 200°F will cure aromatic amines adequately. For the undercured units, a postcure of 150°F for an hour will "fix" them.

H. Aromatic Amine-Cured Epoxy Cures "Wrong"

Check the hardener to see if it has crystallized. This will cause a partial separation of the two constituent materials and can cause a "wrong" ratio, as they each have very different ratios. Reheat at 125 to 150°F to completely reliquefy, remembering, of course, to use an oven vented to the outside.

I. Urethane Expands During Cure

A good guess here is that the reactor portion has been contaminated by moisture. The isocyanate group will preferentially react with any available moisture, releasing carbon dioxide, a gas. Section a piece of the casting and examine under magnification to see if small voids are present. The solution is to keep both components very dry, backflushing cans with dry nitrogen when they are opened. Try heating the reactor portion in a 200°F oven to see if the water will leave. Otherwise, the material is a throwaway.

J. Silicones: Special Problems

RTV silicones are absolved of some of the epoxy and urethane problems, but have a few of their own.

1. Silicone-Cured Except Near Component or Leads

The most likely cause of sticky, semicured, or uncured areas against leads, components, or other materials is suggestive of cure inhibition, usually a phenomenon of addition cure silicones (see Chapter 3). The only solution is to dig out the uncured material, clean up the lead, component, and so on, with solvent, then apply a barrier coat, such as a clear acrylic spray. This uncured area will probably not cure further, even with the use of heat. Materials that can cause cure inhibition are heat-cured rubbers, plasticized vinyl wire insulation, residual flux, amine-cured epoxies, and condensation-cured silicones. The use of heat during cure can eliminate or reduce inhibition, but remember, heat can cause other problems related to silicones' high coefficient of thermal expansion. The best solution is to change the offending inhibiting material. If the barrier coat is not effective, and it is not always, consider changing to a condensation-cure silicone, which is relatively immune to inhibition.

A last *happy* thought on this subject. Unlike epoxies or ure-thanes, semicured silicones have good insulation resistance and di-electric strength.

2. Slow-Through Cure (Firm on Top, Doughy
    and Mushy Toward Bottom)

This is a characteristic of condensation-cure silicones. Two things can cause this:

1. Low temperatures (below 60°F) and low humidity cause such slowing. The potted units should be cured in a warmer area, 75 to 100°F. If units are cured in a closed-in area, a dish of water will supply some beneficial hu-midity. Storing the uncured silicone at 70°F or higher will also help push the cure along in cool conditions.
2. Excessive evacuation of condensation-cure silicones prior to potting will also cause this. The vacuum can actually distill out some of the essential cure ingredients. Normal-ly, 5 min of vacuum is sufficient. Leaving it under vacuum much longer than 5 min is injurious to the RTV and contributes little, if anything, to better air removal. Many of those bubbles you see toward the end of a longer vacuum period may be cure ingredients distilling out.

The above presupposes, of course, that the right catalyst in the right amount has been used. Low catalyst levels will also cause very slow cure, as will temperatures below 70°F. At 50°F silicones cure extremely slowly. A minimum temperature of 70°F is re-quired, and better still, 75 to 90°F.

3. Bulge in a Potted Unit

If the bulge is punctured and a gas escapes, it is probably hydrogen gas, H_2. Addition-cure silicones, when exposed to acids or bases, can generate H_2 in the area of exposure. It tends to accumulate in the interior of the silicone, forming a bladder that distends. This phenomenon is a will-'o-the-wisp and very unpredictable, occurring in, say, 3 out of 10 units today, none tomorrow, then 6 out of 15 the next day. Reexamine cleaning procedures is the best advice if it is encountered.

A condensation-cure silicone will also bulge if it is heat accelerated at too high a temperature. The larger the unit, the more likely this is to occur. Keep temperature below 150°F in small units, less than ½ in. in depth, and to 125°F for larger units.

III. CLASS II MATERIALS

A. Mold Not Completely Filled

This is a problem with many potential causes.*

1. Compression molding
   a. Is the charge size adequate to fill the cavity? Check the *volume* of the charge against the mold cavity volume.
   b. Split the charge into different parts of the mold to help flow.
   c. Lower the mold temperature and increase the time and pressure to encourage increased flow.
   d. Preheat the charge (easier if it is a preform) to promote more immediate flow.
   e. If all else fails, perhaps a higher-flow-rate material is in order.
2. Transfer molding and injection molding
   a. Check to see that the vents are adequate and not partially blocked with flash.
   b. Check the shim stock in the parting line.
   c. Check the mold temperature. If it is too high, the charge will set up before fill is complete.
   d. Overage material can give poor flow.
   e. An old mold can trap gas. Preheating the mold releases absorbed gas and moisture.

---

*For more information, see Ref. 2.

  f. If using a new mold, the gate size, location, and type should be checked.

  g. If the charge or preform is preheated, it may be too hot, causing premature cure.

B. Nonuniform Part Strength

 1. *Compression molding.* This is seldom a compression problem.

 2. *Transfer and injection molding.* An appreciable strength difference in one direction is probably due to orientation of the fiber reinforcement during the molten stage as it passes through a gate. Mold design determines this and has to be corrected.

C. Part Warping After Cure

 1. *Compression molding.* Not a common problem.

 2. *Transfer and injection molding.* This is usually caused by an uneven rate of shrinkage in the flow direction as opposed to material at right angles to it. This is caused by part geometry and location of gates, which cause orientation of fibrous fillers.

Also check uniformity of the mold heating. Hot spots can cause early gelation, and therefore stress, just as in class I materials.

D. Blistering in Thick Sections

Several treatments can help here.

 1. Preheat the material longer, or if it is up to the design temperature, at a higher temperature.

 2. Increase clamp pressure.

 3. Lower the mold temperature.

E. Mold Sticking (Charge Sticks to Mold)

Several things can help here.

1. Raise the mold temperature to speed up the cure. The part may be coming out partially cured.
2. Preheat the material longer to purge out moisture.
3. Clean the mold. Residue will stick to fresh material.
4. Check the mold surface. If it is not shiny, polishing may be in order.

F. Differential Shrinkage

1. *Compression molding.* Not usually a problem.
2. *Transfer or injection molding.* Some materials shrink at a different rate in the flow direction than at right angles to it, particularly fiber-filled materials. Geometry of the mold, gate location, and size should be checked as the fibers are being oriented excessively.

IV. CLASS III MATERIALS

A. Bubbles Under Components on the Board

1. *Dipped board.* If the board is dipped, the immersion speed of the board is probably too fast. It does not allow time for the coating to displace the air around and under a component, but rather, overlaps and traps it. Once trapped like this, bubbles become quite "adhesive" and tend to cling to the board or component.
2. *Sprayed board.* If the coating is being sprayed, there are two potential causes:
   a. Air pressure may be too high, which can blow the coating off as it lands and trap air pockets in "blind alleys" under components. Drop the pressure 5 to 10 lb and try on a fresh board.

b.   The nozzle may be too close to the board, giving the same effect as above. Back off 3 to 4 in. and try again.

3.   *Special urethane problem.* If a urethane coating is being used, there is one other potential source for bubbles. Because of their moisture sensitivity, urethanes will react with any moisture on the board, so be sure that it is thoroughly dry. These bubbles will appear anywhere there is moisture, not just around components.

B.   Coating Thickens as More Boards Are Dipped

There can be two different phenomena involved in this case. One, solvent loss, can involve all the solvent-thinned systems; and the other, only the reactive types, which would exclude acrylics.

1.   Nonreactive coatings do not cure; they simply lose solvent until the coating is dry. Acrylics are this way. As solvent is lost from the reservoir, the residual material gets progressively thicker and a dipped board will, of course, pick up more. The solution is to monitor the reservoir for viscosity or percentage of solids and add solvent as needed. Partially covering the reservoir will retard solvent loss, as will floating hollow glass spheres on the surface.

2.   Reactive coatings can also exhibit this phenomenon, and the solution is the same. However, progressive reaction of the two components will also cause a viscosity rise independent of solvent loss. There are several things that will help in this case:

a.   Use a slower-curing system.

b.   Dilute with more solvent. This will lower the viscosity, but ultimately will result in too thin a final coating, as it lowers the solids.

c.   Use a smaller reservoir and add freshly mixed material more often.

d.   Cool the reservoir to slow the rate of reaction and viscosity rise.

C. Coating Does Not Harden (Remains Leathery)

Also two possible causes: one, solvent retention, which can afflict both reactive and nonreactive types, and the other, a wrong reactor-to-base ratio, which affects only reactive materials.

1. Nonreactive coatings
   a. Solvent retention results if too much coating is applied in one coat. The top surface of the film loses solvents, skins over, and traps the remaining solvents under it. Apply less per coat and more coats if necessary.
   b. Add solvent to the reservoir to lower the total resin solids and, therefore, to lower the amount picked up per dip or retained when sprayed.
   c. A slower evaporating strong (high-solvency) solvent can be added, which keeps the film soft and "open" until the other, faster solvents can evaporate out. This slightly retards the overall dry time but discourages solvent entrapment. This should be done with your supplier's advice.
2. Reactive systems
   a. See items a to c in Section IV.C, as they all apply here as well.
   b. Check to see that the correct ratio of base to reactor was used. If this ratio is off, incomplete cure can result.

D. Uneven Coating on the Board

1. This is often a characteristic of brush coating. The only solution is to "brush better" or use another method. Several very light coats will give a more uniform thickness than one or two heavy ones.
2. Very low coating solids resulting in low viscosity will cause excessive runoff with some puddling. Try turning

the coated board during the drying period to aid in a more uniform distribution of the coating, then increase the coating solids level.

V. PARYLENE

The Parylene process is a turnkey operation, purchased or leased. Call the equipment supplier when problems are encountered.

REFERENCES

1. Buchi and Flynn, Pressure Gelation Process, Reinforced Plastics Composites Institute, 1974. Society of the Plastics Industry, New York.
2. *Thermoset Molding*, Society of the Plastics Industry, New York.

8
Decapsulation

I. INTRODUCTION

Among the earliest considerations during the material selection process is an answer to the question: Must an encapsulated unit be repaired after the encapsulation step if found to be defective, or is it a throwaway? If encapsulant removal to effect repairs is necessary on a production basis, this information is critical input during the material selection step.

There are two ways to remove encapsulants:

1. *Mechanically*: cut, grind, peel, and so on
2. *Chemically*: strip or dissolve away

Of these two, only the first has utility for material removal and unit repair on a production basis, as opposed to failure analysis. The chemical stripping action is not only too slow, but it causes irrepairable damage, obviously something that will strip a thermoset phenolic, DAP, or an epoxy will also attack tapes, coatings, adhesives, and marking inks used on components. But the stripping action is slow. As the stripping action proceeds, the surface is softened, but unless continuously removed, it inhibits further stripper penetration, so the penetration slows to a creep. It has utility only as a part of failure analysis of a defective device.

The exceptions are class I and class II materials, which contain high levels of calcium carbonate filler. Many strippers contain acidic components, such as formic acid, and attack the filler as well as the resin. The continuous bubbling action helps break up the encapsulant. However, tapes, inks, and so on, are still attacked.

Among the many materials that have been studied in the materials chapters, only one, RTV silicone, is truly easily mechanically removable. Soft epoxy and urethane gels are much tougher and adhere much more tenaciously, and although removable, are very difficult to remove on a regular or production basis. None of the thermoset powders are easily removable. Of the conformal coatings, the acrylics come closest, but it is a rather sticky and slow process.

II. CLASS I MATERIALS

A. Three Questions

There are three important questions that must be answered prior to final material selection.

Question 1. When potting or casting devices, the first question to ask is: Must the encapsulant be removed to effect repairs, or, once encapsulated, are reject units throwaways?

If the answer to the question is positive, the encapsulant must be removable. *Then* the choice of liquid potting materials is greatly narrowed—to RTV silicones, or possibly a few very soft epoxy and urethane gels.

Question 2. The second question is now dictated by the answer to Question 1. It is: Will the narrow list of available materials give the economic and functional end result required by the component or device? If the answer here is yes, answer Question 3.

Question 3. RTV silicones, epoxy gels, or urethane gels? RTVs are *by far* the easiest to remove as they cut and pull away easily, and even when used with a primer, the adhesion to components and case is not a serious deterrent to removal. The epoxy and urethane gels, however, are not only much tougher, particularly urethanes, but they adhere well to components and case and are much more difficult to remove mechanically. *However*, RTVs are some two to four times as expensive as epoxies and urethanes—can the device support this increased cost? *But* RTVs offer superior insulation resistance compared to soft epoxies and urethanes—10^{15} versus 10^{12-13} Ω-cm, and at elevated temperatures the gap widens. Is this a drawback, or does it matter in the particular device?

B. Materials Discussion

1. RTV Silicones

They stand alone in their ease of removal for replacement of a defective component and resealing of the unit. A small scalpel or a small jackknife will quickly and easily chip away cured RTV. Even primed RTV is much easier to remove than the softest epoxies and urethanes. Where wires are so fine that any mechanical stress would result in breakage, mechanical removal can be performed down to the critical area; then a selective stripper can be used. However, strippers are messy, hard to contain in the desired area, and *must* be completely removed before attempting to reseal or pot. Failure to completely remove and clean the area can cause a myriad of problems:

Cure inhibition of fresh RTV
Poor adhesion of fresh RTV
Sealing in of a potentially corrosive material
Contamination of components in the area

Potential short-circuit paths

Thermal cycling problems, as strippers are quite volatile

Mechanical removal is much preferred if it can possibly be managed (see Fig. 1).

2. Epoxies

Much, much harder to remove than silicones. With the exception of the gels and a few (semiflexible) dipping materials, epoxies cannot easily be cut into and chipped away. Only a chemical stripper will remove them, and this is usually too slow, too messy, and too damaging, as already discussed. In potted or cast units, the rejects are dumped. It is a powerful incentive to improve quality control prior to the potting step. If the unit will not operate in air (for a final check before potting), it should be filled with an inert dielectric fluid (mineral oil, fluorocarbon, silicone fluid), final tested, the fluid removed, the unit drained and cleaned, and then potted. Although this sounds messy, it sure beats chemical stripping. Recently, inert fluorocarbons have become available for such final testing. They are super pure, inert, drain completely, and leave no residue.

Units Potted in Cans or Cases.   These are usually impractical to remove. The case interferes with removal. It is almost impossible to chip away a hard material such as epoxy, confined to working from only one open end.

Cast Units (No Case).   If in semiflexible or low-softening-point materials, they can be heated (to 125 to 150°F) and chipped away with a screwdriver, chisel, or knife. This is a low-volume technique suited only to a few very large or very expensive units. The unit must be very robust to withstand the physical pulling and peeling forces generated during removal, to say nothing of misdirected hammer and chisel blows. Protective gloves, goggles, and so on, are mandatory here.

Dipped Units.   If in a semiflexible or low-temperature-softening material, the unit can be heated (125 to 150°F) and the encapsulant cut, chipped, and sometimes peeled away. This is similar to the procedure used with the cast unit but somewhat easier. The material exists as a thin, 25- to 150-mil envelope, which lends itself more to peeling than if cast into a block. As above, it is practical only for a few units at a time.

Advice. Because of the difficulty of epoxy removal, prevention is the best medicine. Make sure that the unit works before "locking it up" in epoxy.

3. Urethanes

Urethanes lie between epoxies and silicones in removability, unfortunately, much closer to epoxies than silicones. They are too hard or tough for practical mechanical removal, so end up being treated much like epoxies. As they soften more readily than most epoxies, they lend themselves a little more to heat up and chip, chisel, or peel away, but are by no means easy. However, because of their inherent toughness, they can cause a lot of broken wires and solder joints. Final testing requirements are like those for epoxies. Make sure that it is right before you bury it.

4. Polyesters

Their lower softening points make "picking" the polyesters out of a potted or cast unit a little more practical than with urethanes and epoxies but still not practical on a volume basis. Preheating, as with epoxies and urethanes, is necessary. Polyesters strip easier than either epoxies or urethanes, but this process is not usually practical because of internal damage to the unit.

5. Summary

The bottom line on recovery of class I encapsulated units can be summed up this way: If cost precludes the use of RTV silicones, the time and cost of recovering a reject unit out of epoxies and urethanes must be less than the salvage value of the unit; or make it right before you encapsulate it.

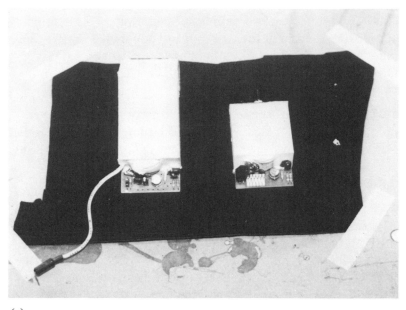

(a)

Fig. 1   (a) Power supplies cast in R. T. V. silicone. (b) Power supply with high voltage section partially "picked away." (c) Area near transformer is picked down to board level making component replacement possible. Photographs courtesy of International High Voltage Electronics, Danbury, Connecticut.

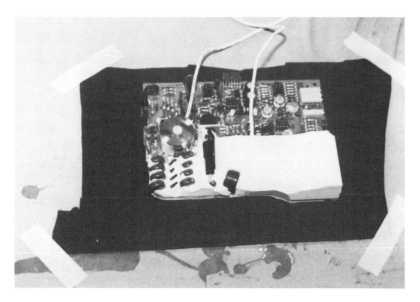

(b)

(c)

III. CLASS II MATERIALS: REPAIRS

None of the class II materials used can be considered removable. Fortunately, the type of device they are used to protect is usually much simpler than that encapsulated in a class I material, so there is less likelihood of a need to remove it. Coils lend themselves more to stripping than a stuffed board, but time and chemical damage are usually still prohibitive. The problem with molded class II materials then becomes more one of repairing poorly or incompletely molded units.

A. Surface Imperfections

Surface pits and small cavities can be repaired and be cosmetically acceptable by the use of some of the original molding powder and a soldering iron. Some of the original powder is pressed into the hole, then softened and smoothed with the soldering iron. The iron is held on until gelation takes place.

B. Larger Imperfections

For larger imperfections, a one- or two-component thixotropic epoxy (of the same color and shade as the molded part) can be "puttied" into the hole. Both room-temperature and heat-cured materials have been used successfully. Time must be provided for cure, of course.

C. Partially Filled Units

Where the fill is only partial and a large portion is missing, the area (on the casting) around the gate is ground away, then a groove or runner is ground across to the unfilled portion. The "partial coil" is then placed again in the mold cavity and a partial shot of

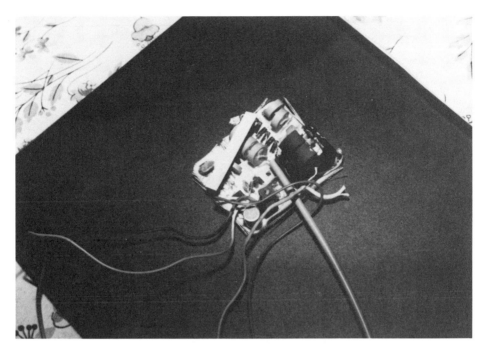

Fig. 2   A chemically-stripped power supply. Note that all markings and coatings have been destroyed and removed. As well, the base G-10 board has swel-
led, warped and delaminated. Pressure-sensitive tapes are also removed and in
most cases remain only as a stringy strip of plastic. The unit has been effectively destroyed.

powder is used to refill it. Note: The charge size must be carefully
monitored.

IV. CLASS III MATERIALS

Conformal coatings, as a class, are less difficult to remove than the
other two classes of materials (except RTV silicones), although
there is a great variation in degree of difficulty of removal between

individual conformal coatings. The fact that they exist only as thin coatings makes all forms of attack more feasible.

A. Acrylics

Acrylics are the easiest of the conformal coatings to remove.*

1. Solvent Removal

Since the acrylic film is formed solely by solvent evaporation, it can be redissolved by application of the same solvent. This can be done by soaking the whole component or board in the solvent for complete removal. For "spot" removal, to permit a component change on a board, a cloth soaked in solvent is applied to the area until the coating is removed. Most chlorinated solvents will effectively remove acrylic coatings. However, note should be made of the fact that the base PCB may be attacked by the solvent selected, as may be component coatings, marking inks, tapes, and so on. Test first. Chemical strippers are not necessary for acrylic removal.

2. Burn-Through

Acrylics can also be "solder removed" simply by melting through the coating with a hot iron and removing the component. The molten solder helps removal of the coating. Very little darkening of the acrylic results and good cosmetic touch-ups are possible.

B. Urethanes

Urethanes are more difficult to remove than acrylics.

1. Chemical Stripping

Urethanes can be removed by selective strippers, which, hopefully, selectively dissolve the coating while not attacking the base board or coated or marked component. However, great care must be

---

*For more information, see Ref. 1.

exercised to avoid attacking the board or corroding metal surfaces, and once removal is effected, to be sure that all stripper residue has been removed before attempting to recoat the unit. Residual stripper can cause a soft spot in the recoated film and become a locked-in source of corrosion and short-circuit paths.

2. Burn-Through

A second removal method involves use of a soldering iron to solder through the coating. As the solder melts, it removes the coating with it, most being pushed away with the flowing solder with some vaporizing. The ease of solder removal of urethanes is dependent on the film thickness. Up to 4 mils it goes fairly easily but much thicker than that is more difficult. Cosmetics are also a consideration, as the heat will cause discoloration of the surrounding area, which persists to some degree after touch-up. When resealing with fresh urethane, rough parts and edges should be sanded down with emery paper for the sake of cosmetics.

C. Epoxies

Epoxies are somewhat more difficult to remove than urethanes, as they have better chemical and solvent resistance.

1. Chemical Stripping

A stripper capable of removing the epoxy coating also rapidly attacks the PC board and other components, as well as potentially corroding some metals. If a stripper is used, it must be selectively and carefully used to prevent the other problems. As with urethanes, it must be thoroughly removed before attempting to recoat.

2. Burn-Through

In view of the drawbacks and dangers of stripping, heat soldering through is probably the best removal technique. This, too, is somewhat more difficult than with acrylics or urethanes but can be done, with, however, some additional evidence being left behind.

Discoloration from the heat and irregularities from softened old coating will be noticeable. The irregularities should be removed with emery paper and/or a scalpel before touching up with fresh epoxy.

D. Silicones

Silicones are more difficult to remove even than epoxies, as most solvents and strippers do not effectively remove them without attacking the board and components. Soldering, through, is also not effective because of silicones' excellent heat resistance; silicones do not soften with heat nearly as much as the others and do not vaporize. This leaves only mechanical means; cutting the film and peeling and scraping it away to expose the targeted place or component.

E. A Third Removal Technique (2)

Recently, information was published on a third possibility in conformal coating removal. This is abrasive blasting. A finely collimated stream of abrasive particles cuts through and removes the coating only in the areas on which it is focused, without harming surrounding areas. Where only a small amount of coating needs to be removed, such as on a solder joint, it could be done very quickly and leave no harmful residue or heat discoloration marks. It must be carefully aimed and manipulated with a carefully regulated air supply. Care must be taken that the base board or nearby components are not cut into.

F. Resealing

Now that the coating is removed and the component replaced, it is time to reseal or recoat the reworked area. When chemical strippers have been used, all residue *must* be completely removed, as even traces will adversely affect adhesion of the repair coating and can leave a "built-in" corrosion problem if locked in under the new coating. After thorough cleaning, the repaired area should be sanded with a fine grit emery paper to remove any pieces of previous coating and to bevel the edges of the old coating. This will help the repair coating assume a smooth, more cosmetic surface appearance. Sanding will also improve adhesion of the repair coating.

The same coating that was originally applied need not necessarily be the one used on the repaired area. Epoxy is often used as the repair coating because of its excellent adhesion to the other coatings and its ease of use.

V. PARYLENE

Parylene is the most difficult conformal coating to remove. Strippers and mechanical methods are not effective. The new abrasive blasting offers the best technique for small area removal. Plasma etching has been used in failure analysis but would be impractical on a production repair basis. Do it right before you coat.

Carbide revealed in late 1985 (3) a new, more removable Parylene—Parylene E—giving almost equivalent electrical properties to Parylene N, C, and D. Parylene E is a copolymer of E-type monomer with C type. The resultant film is removable with toluene, hexane, xylene, and several other solvents commonly found in electronic facilities.

REFERENCES

1. J. Waryold, "How to Use Conformal Coatings Efficiently," Columbia Chase Corp., Humiseal Division, Woodside, N.Y.
2. Removing the unremovable with abrasive blasting, *Electron. Packag. Prod.*, May 1984.
3. Solvent strippable parylene coating, *S.A.M.P.E. Quart.*, Oct. 1985.

Index

A

Acrylics, 65, 66–68, 165–166, 216
Anhydrides, 30–32
    three major anhydrides, 31–32
Amines, 21, 22–28, 45
    adducts, 25
    aromatic, 28–30, 198
        cures wrong, 198
        epoxy hard, brittle, 198
    copolymers–amidoamines, 26–27
    copolymers–polyamides, 25–26
    general formula and description, 25
    property summary, 27
    special amines, 28

B

Blistering in thick sections, 202
Bubbles under components, 203–204

C

Class I materials, 19–56, 78–111, 137–160, 177–190, 191–201, 207–213
Class II materials, 57–65, 112–126, 160–164, 201–203, 213–215
Class III materials, 65–75, 126–132, 164–172, 203–206, 215–219
Cleaning, 180–182

Coating doesn't harden, 205
Coating thickens (in reservoir), 204
Compatibility, 178–180
Corona ground shield, 189–190
Cures too fast, 193
Cures too slowly, 193–194
Curing agents, 21–34, 160
    catalytic, 32–33
    other–heat cure, 33–34
    overview, 21–22
    safe handling, 160

D

Decapsulation, summary Class I ma-
    terials, 211
Depolymerized rubber (DPR), 51–52
Diallylphthallate (DAP), 59, 162–
    163
Differential shrinkage, 203

E

Elevated temperature cure, 189
Encapsulation process (High voltage
    assemblies), 184
Encapsulating, 184–188
Epoxies, 19–27, 34–36, 60–61, 63–
    65, 70–71, 142–144, 161–162,
    167–169, 210–211, 217–218
    cresol novolacs, 63–65
    cycloaliphatic, 34–35
    general formula and descriptions,
      19–20
    novolacs, 63–65
    other epoxies, 34–36
    recovery of components from,
      210–211, 217–218
    rubber modified, 35–36

F

Fillers, 36–37, 42, 46, 49, 151–157
    in epoxies, 36–37
    in polyesters, 49

[Fillers]
    in silicones, 42
    in urethanes, 46

G

Guidelines for high voltage encapsu-
    lation, 190

H

High shrinkage, 196–198

I

Inconsistant cure, 195–196

L

Large imperfections, 214

M

Material/assembly degas, 186–187
Material evacuation, 185–186
Materials discussion, 209–214
Materials of the future, 171
Methods of application, 86–132
    Class I materials, 86–112
      Buttering-on, 102
      Dipping, 97, 102
      Hand mixing, 86–87
      Impregnation, 103, 106
      Mechanical potting, 91–97
      Other methods, 107–112
      Pressure, 107
      Vacuum, 87–90, 90, 106–107
      Vacuum degas, 87–90
      Vacuum pot, 90
    Class II materials, 112–126
      Compression molding, 113–
        116
      Electrostatic spray, 124–126

[Methods of application]
    Fluid bed, 121-126
    Injection molding, 120-121
    Transfer molding, 116-120
  Class III materials, 126-132
    Brush coating, 129
    Dip coating, 126
    Flow coating, 129
    Parylene, 129-132
    Spray coating, 126-129
Mold not filled, 201
Mold releases, 78-80
Mold sticking, 203
Molds vs shells, 80-86

N

No cure, 192-193
Non-uniform part strength, 202

P

Part warping after cure, 202
Partially filled unit, 214-215
Parylene, 73-74, 129-132, 170-171, 206, 219
Phenolics, 61-62, 162
Polybutadiene, 49-51
Polyesters, 47-49, 62-63 (BMC), 149-151, 163-164 (BMC), 211
  General formula and description, 47
Polysulfide polymers, 52
Preliminaries to encapsulation, 78-86
Priming, 182-183

R

Resealing, 218-219
Room temperature curing materials, 188-189

S

Safe handling
  class I materials, 159-160
  class II materials, 164
  class III materials, 172
Silicones, 37-42, 71-72, 144-147, 169-170, 199-201, 209-210, 218
  general formula and description, 37-38
  cure mechanisms, 38-39
Simple encapsulation, 185
Soft/hard spots in cured material, 194-195
Summary
  class I materials, 157-159
  class II materials, 65
  class III materials, 72
Surface imperfections, 214

T

Third removal technique, 218
Three questions (Material Selection), 208-209

U

Uneven coating on board, 205-206
Urethanes, 42-46, 68-69, 147-149, 166-167, 198, 211, 216-217
  general formula and description, 42-44
  isocyanate crosslinkers, 45
  isocyanate sources, 44

V

Vacuum encapsulation, 187-188